KUHN R. S.

Fundamentals of Drilling
—Technology and Economics

Fundamentals

of Drilling

—Technology and Economics

John L. Kennedy

PennWell Publishing Company
Tulsa, Oklahoma

This book is dedicated to the following persons, listed in the order of their initial influence on my life:

> Russell and Wilma
> Barbara
> Patty
> Jane
> Anne

Copyright © 1983 by
PennWell Publishing Company
1421 South Sheridan Road/P.O. Box 1260
Tulsa, Oklahoma 74101

Library of Congress cataloging in publication data

Kennedy, John L.
  Fundamentals of drilling—technology and economics.

  Includes bibliographical references.
  1. Oil well drilling. 2. Gas well drilling. I. Title.
TN871.2.K39 1983   622'.338   82-13260
ISBN 0-87814-200-2

All rights reserved. No part of this book may be reproduced, stored in a retrieval system, or transcribed in any form or by any means, electronic or mechanical, including photocopying and recording, without the prior written permission of the publisher.

1 2 3 4 5   87 86 85 84 83

Contents

Preface   vii

1. **The oil and gas well drilling industry**   1
   A brief history 1
   Offshore drilling 5
   Technology for all environments 10
   Types of wells 11
   The parties involved 13
   Key industry statistics 16

2. **Types of rigs**   29
   Land and offshore 30
   Rotary rigs 32
   Source of power 35
   Offshore rig types 36
   Moving rigs 45
   Workover rigs 47

3. **Major rig components**   50
   Hoisting 52
   Rotating 57
   Fluid handling 59
   Drill string 62
   Other equipment 65
   System efficiency 65

4. **Drilling bits**   69
   Types of bits 70
   General considerations 77
   Bit selection 81

5. **Drilling fluids** 83
   Circulation path 83
   Purpose of drilling fluids, 85
   Drilling fluid can solve problems 90
   Types of drilling fluids 93
   Mud properties 96
   Drilling fluid treating and monitoring equipment 98
   Advances in drilling fluids 100

✓ 6. **How the hole is drilled** 105
   Well planning 106
   Moving in, rigging up 118
   Drilling the well 119
   Offshore floating drilling 131

7. **Directional drilling** 135
   Directional drilling applications 136
   Directional drilling methods and tools 144
   Some problems 150

8. **Well control and safety** 152
   Consequences of blowouts 152
   When blowouts are most likely 155
   Flow of formation fluids 157
   How kicks are handled 160
   Blowout prevention equipment 162
   After the well blows out 167

9. **Completing the well** 172
   Logging 173
   Drill-stem testing 178
   Perforating 179
   Sand consolidation 180
   Stimulation 181
   Producing equipment 183
   Testing after completion 187

10. **Today's and tomorrow's technology** 190
    New bits 191
    Measurement while drilling 193
    Downhole motors 196
    Deep drilling 198
    Training 199
    Data gathering and analysis 201
    Research needs 203

Index 209

Preface

DRILLING an oil or gas well is a complex operation. Though similar methods and equipment can be used in many wells, literally each well is different. Consequently, oil and gas well drilling has little in common with typical industrial production-line operations.

This overview of the common techniques, the equipment, and some of the problems involved in drilling oil and gas wells is designed to familiarize those persons not directly involved in drilling operations with the industry and its capability. In addition, it explains some common industry terms and relationships and offers perspective on the tremendous cost of oil and gas well drilling. The book also indicates the diverse skills that must be combined to drill a single well. Extrapolate this complexity to the tens of thousands of wells drilled around the world each year—some under the most difficult conditions imaginable—and the remarkable accomplishments of the industry will become apparent.

This book is not a drilling operations manual or a source of well design information. If design and operations techniques were detailed for many individual aspects of drilling, an entire book would be required for each subject. Instead, the text provides information that will help clarify basic drilling procedures without complicating the subject with data only a drilling engineer could profit from.

Much has been written on all phases of the drilling operation, and design and operating detail is available on almost any subject discussed in the following pages. A sampling of these sources is contained in the references following each chapter.

It is hoped that, in addition to answering many questions, this book will spark an interest in learning more about oil and gas well drilling.

1 The Oil and Gas Well Drilling Industry

OIL and gas well drilling has continued to yield to the march of technology with sophisticated equipment and computer analysis of drilling variables an accepted part of today's drilling operation. But there is still a considerable amount of art involved in drilling a hole to a depth of more than four miles and then installing equipment in it while not being able to see what is being done. Drilling crews perform complex operations thousands of feet down to set tools or to remove them from the hole, guided only by their experience in analyzing a variety of gauges and conditions at the surface. Even so, today's drilling technology is the result of only a few dramatic innovations; continued steady evolution is responsible for most of the advance made in the last 100 years.

It is impossible, of course, to say precisely what new techniques and equipment will become a part of routine drilling in the next two decades. But it is likely that most will still be the product of evolution rather than the result of a dramatic new direction. As this evolution continues, the drilling operation will involve more and more applied science. But, as will be apparent after reading this basic outline of oil and gas well drilling, successful drilling will never be possible without at least a little art. Performing a critical operation with a tool on the end of a dangling string of drill pipe 20,000 ft long in a hole that may be no larger than a residential air-conditioning duct will continue to demand more than just skill.

A brief history

Almost as far back in history as one wants to go, there is evidence that holes were dug into the earth for a variety of purposes using the tools available at the time. Though not the earliest example, brine wells apparently were drilled in China around 600 B.C.[1] One analysis of records estimates that as early as 1200 A.D. wells may have been drilled as deep as 1,500 ft. Much more detailed accounts of these ancient drilling efforts are available.[2]

But a good reference point for beginning a brief look at the development of modern drilling methods is the Drake well, drilled by Colonel Edwin Drake in Pennsylvania in 1859. Though not a deep well—total depth was 69 ft—and not a highly productive one at about 20–30 b/d by pump, it is significant for two reasons. First, it is considered by many to be the first commercial oil well. Second, it was drilled with equipment that would become the standard for many years to come: the cable tool rig.

Cable tools. Cable tool rigs make hole by raising the bit with a system of wheels and cables and then dropping it, punching the hole deeper. These rigs were the workhorses of the oil and gas drilling industry for almost a century. They drilled most of the wells and found large reserves of oil and gas. But they began to be replaced with rotary drilling rigs in the early part of of the 20th century, and after World War II continued to give ground to the rotary rig.

Rotary rigs. The rotary rig drills the bulk of oil and gas wells today. A rotary drilling machine was patented in 1845, but its most significant commercial debut was in 1901 when it was used to complete the Spindletop well near Beaumont, Texas. The rig was brought in on the Spindletop well after several attempts to drill with a cable tool rig were unsuccessful because of running quicksand. With the rotary drilling equipment, this troublesome zone could be drilled and isolated with casing.

Rotary rigs make hole with a boring action rather than punching a hole as the cable tool rig does. A bit rotates while it is in constant contact with the rock at the bottom of the hole. Part of the weight of the drill pipe above the bit rests on the bit while it is rotating to force it into the rock as it turns. Fluid pumped down the drill pipe and back to the surface removes rock cuttings from the hole.

The share of oil and gas well drilling done by rotary drilling equipment has continued to increase steadily since the 1940s. Today, rotary rigs (Fig. 1–1) drill most oil and gas wells.

The basic concept of rotary drilling—rotating a bit on the bottom of the hole with a length of drill pipe through which fluid is circulated to remove cuttings—has not changed significantly in more than 75 years.

Fig. 1–1 Modern land drilling rig. (courtesy Oil & Gas Journal)

But there have been many improvements in the equipment comprising these rigs. These improvements have brought greater efficiency, greater depth capability, and more control over hole conditions and reservoir fluids. And new tools have been developed to supplement the basic rotary drilling machinery.

One of the most significant advances was the rolling cutter rock bit. It was designed, built, patented, and used by Howard Hughes in 1909. Early bits of this type underwent considerable revision, but the roller bits used today are quite similar to those used a half century ago. The concept is the same, but vast improvements have been made in design, metallurgy, and components. Today's bits last many times longer than those early bits did, and the variety of available types makes the roller cone bit applicable to almost all formations.

In recent years, development has focused on making bits match the characteristics of the rock they will be drilling. Many modern bits are designed to drill a variety of formation types.

Other equipment used for the rotary drilling operation has been greatly improved. Circulating systems have greater capacities and can be much more precisely controlled. Drilling fluids have been formulated using complex chemistry to combat specific downhole problems. The chemical and physical properties may even be adjusted several times while drilling a single well. Drill pipe metallurgy has been improved to combat corrosion and to withstand the stresses resulting from extreme temperatures, pressures, and depths. Pipe handling tools and downhole equipment have been developed to meet specific needs. For example, in offshore drilling where handling heavy equipment on a floating vessel is dangerous, automatic pipe handling systems have been developed. Downhole tools to permit cementing casing and performing other jobs in extremely deep wells are other examples of this development.

But rotating the bit while it is in contact with the rock at the bottom of the hole and circulating fluid through the drill string to remove cuttings is still the way oil and gas wells are drilled today.

Other approaches. This is not to say considerable time and money have not been spent in an attempt to find a better way to drill. Several changes in the basic concept of rotary drilling have been studied, and some have been field-tested. To date, however, they have not been competitive with the conventional rotary drilling technique.

Two of these techniques, high-pressure water jetting and abrasive jetting, still use mechanical energy to remove rock from the bottom of the hole. However, they are significant departures from the conventional rotary method because of the specially designed equipment needed. Both of these approaches have been field-tested with some success, but neither is in routine drilling use.[3]

Other approaches to drilling oil and gas wells have been studied but are still not in commercial use. These include:

1. electric arc and plasma drills that melt, spall, or thermally degrade the rock
2. electron beam drills that melt and vaporize the rock
3. explosive drills that shatter the rock formation
4. laser drills that melt or vaporize the rock

Even more exotic ways to remove rock have been studied, but none can yet compete with the conventional rotary drilling process in 'round-the-clock drilling under tough field conditions. Many changes will be made in drilling methods in the next two decades, but there is little evidence that any conceptual change in today's rotary drilling technique will make a significant contribution in the near future.

Offshore drilling

Although the method used to drill a well in the ocean floor is similar to that used to drill a well on land, special equipment is needed to support the rig above the water. Because of this, offshore drilling has become almost a separate industry with an impressive, though short, history of its own.

Wells were drilled over water from a pier at Santa Barbara, California, as early as 1897. In 1911, a steam-powered rotary rig drilled from a wooden platform in Caddo Lake, Louisiana. And in 1933, a drilling rig mounted on a barge drilled in Lake Pelto, Louisiana. The modern offshore drilling industry got its real start in 1947 when the first well was drilled out of sight of land by Kerr-McGee Corp. in Ship Shoal Block 32 in the Gulf of Mexico. In the few years that followed, technology developed rapidly. In 1955, drilling was performed from a drillship, a ship-shaped vessel on which a rotary rig was mounted.

The capability of the industry to explore and develop oil and gas reserves in the oceans of the world was apparent by 1957 when a well was drilled in 100 ft of water. The technology needed to drill in that water depth pales beside today's deep-water drilling technology. The equipment used a short 25 years ago seems primitive compared with the sophisticated offshore drilling units of today (Fig. 1–2).

In a mere quarter of a century, the offshore drilling industry has developed drilling vessels and support equipment that make it possible to drill in water depths of several thousand feet. In 1979, for example, Texaco Canada Resources Ltd. drilled an offshore well in 4,875 ft of water off Newfoundland.[4] That record was expected to be challenged in 1982 in the Mediterranean.

Limited experimental drilling operations have been conducted in much greater water depths. In 1970, the Deep Sea Drilling Project reentered a hole drilled in 13,000 ft of water in the Caribbean.[5] A hole was drilled to 2,300 ft below the ocean floor with the first bit; then the hole was located and reentered and another 200 ft was drilled with the second bit.

Although commercial oil and gas well drilling has not been done in water nearly this deep, this achievement was significant. It demon-

6 FUNDAMENTALS OF DRILLING

strated the ability to reenter a hole in very deep water. Reentry is a key to successful deep-water drilling.

The water depths in which oil and gas can be produced are less than the depths in which the industry can drill. At present, technology is not available to produce from water depths of 5,000 ft, for instance. But that technology could be developed from existing hardware if large hydrocarbon reserves are expected and if the economics are favorable.

Production technology is constantly being developed for deeper water and for more severe environments. For example, fixed production platforms, still the industry's preferred method of producing offshore fields, have been installed in 850 ft of water offshore California and in 1,025 ft of water in the Gulf of Mexico. Another fixed platform was being built in 1981 for installation in the Gulf of Mexico in 935 ft of water. These and other fixed platforms for more modest water depths support oil and gas producing equipment above the water's surface. Platforms that reach above the water's surface but are not rigidly fixed to the ocean floor are also under development. This compliant approach to platform design includes tension-leg platform structures. The maximum water depth in

Fig. 1-2 Modern offshore mobile drilling rig. (courtesy Oil & Gas Journal)

which fixed platforms can be installed is estimated to be 1,200–1,500 ft, while the tension-leg platform might extend platform capability to water depths of several thousand feet.

Subsea completions. There is yet another approach to producing oil and gas in deep water: the subsea completion. When a well is completed subsea, the wellhead and associated valves are installed on the ocean floor and the well is connected to production facilities with a submarine pipeline.

Subsea completion systems range from a single wellhead to complex ocean floor templates through which a number of wells would be drilled directionally and connected to the template. More complex systems have been proposed that would let personnel service wells and equipment on the ocean floor in a dry environment.

Most operating subsea completions, however, are single wells located in water depths up to about 400 ft. A few have been installed in water as deep as 600 ft. About 140 subsea completions had been installed around the world by 1978.[6]

Equipment development. Early offshore drilling was done by mounting land drilling equipment on some sort of makeshift platform to support it over the water. To operate in the water depths the industry needed to reach, platforms and vessels designed specifically for supporting offshore drilling operations had to be developed.

From these early makeshift platforms, equipment evolved into sophisticated steel jackup rigs with legs that can support the platform above the water in depths to 300 ft and more. The modern semisubmersible rig, which does not have to depend on legs resting on the ocean bottom for support, can drill in water depths up to several thousand feet. Special drillships were also developed for deep-water drilling in remote areas. These vessels have large storage capacities for pipe and supplies; some are self-propelled to speed the trip between locations and to eliminate the need for tow vessels.

Another example of the sophistication reached by the offshore drilling industry in its short history is the dynamically positioned drilling vessel. Developed for very deep water and special operations, such vessels can maintain a position over the well site while drilling without using anchors. A series of thrusters maintains position by responding to the commands of a position-monitoring system. A computer links the two main components of the dynamic-positioning system—monitoring equipment and thrusters. Such vessels are few because this capability is not often needed and holding the drilling vessel in position with a conventional anchoring system is a less-expensive way to keep the vessel on location.

Continued growth. Tremendous reserves of oil and gas have been found offshore around the world.[7] Oil produced from offshore fields accounted for about 20% of world production in 1981, averaging about 12 million b/d. Worldwide offshore production of natural gas is about 25 billion cfd, roughly half of which is produced from U.S. offshore fields.

Several thousand wells are drilled offshore each year, and the mobile offshore drilling rig fleet (Fig. 1-3) included almost 600 operating units in early 1982. Another 128 new mobile offshore drilling units were under construction at that time.[8] In addition to mobile offshore drilling units, the industry had about 380 fixed platform rigs located around the world at the end of 1981.[9] Most forecasters feel the fleet will expand rapidly in the next decade as more offshore areas become available for exploration. Modest but steady growth is also expected in offshore oil and gas production during this period.

Development of offshore drilling equipment has been expensive. The cost of a modern deep-water floating drilling rig can be more than $100 million today, and the cost of drilling and completing a single deep-water offshore exploration well can be tens of millions of dollars. Besides the drilling rig itself, which remains on the well site until drilling is complete, the industry had to develop support equipment and services: diving, transportation for crews and equipment, and others. For example, a large fleet of tug/supply and crew/utility vessels now exists to serve offshore drilling and production operations. One estimate puts this support fleet at about 3,500 vessels worldwide.[10]

An industry expert says each mobile offshore rig must be serviced full time by about two offshore supply vessels, and a typical fixed production platform needs the services of one vessel. A sizable fleet of helicopters also supports offshore drilling and production operations, primarily transporting personnel between the platforms and shore bases.

To bring the oil and gas found offshore to market required the development of an offshore pipeline construction capability. Like the offshore drilling industry, offshore pipeline construction is highly specialized. Pipe-laying barges and associated equipment used to build offshore pipelines are equally as sophisticated and expensive as offshore drilling equipment.

Drilling equipment is the same. Whether drilling on land or offshore, the basic equipment used to make hole is virtually the same. Special equipment has been developed to handle conditions that are

THE OIL AND GAS WELL DRILLING INDUSTRY 9

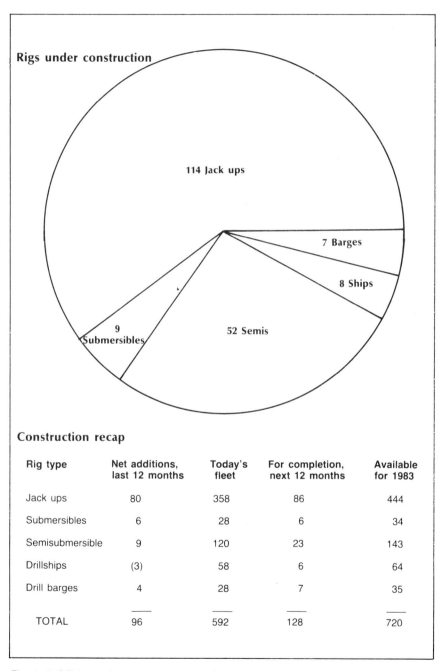

Fig. 1–3 Offshore rig types. (courtesy Oil & Gas Journal, ref. 8)

peculiar to some types of offshore drilling, such as tools to compensate for the constant motion of a floating drilling vessel. These are designed to keep the bit in constant contact with the rock at the bottom of the hole during the up-and-down motion of the rig. Other equipment has been developed to make the handling of pipe safer on floating rigs.

Aside from these modifications for the special conditions encountered in a marine environment, the hole is still drilled the same way as a hole would be drilled on land. The rotary rig, which is mounted on the drilling vessel, has the same basic equipment as a land rig—rotary table, drawworks, mud pumps, mast. And the rig's function is the same: To rotate the bit in constant contact with the rock at the bottom of the hole and to remove rock cuttings by circulating drilling fluid down through the drill pipe and back to the surface.

The difference is that in offshore drilling this conventional rig and its equipment must be supported on a marine vessel that usually costs many times more than the rig itself. And supply and service for the drilling operation is complicated by the marine environment.

In addition to land and offshore, there are many other ways to classify wells—exploration vs. development, for example—but the same basic type of drilling rig is used to drill most wells. Its size may vary; its support may differ, as in the case of offshore drilling; and it may have special auxiliary equipment for a special job. The basic components, however, are the same, regardless of the type of well being drilled or its location.

Technology for all environments

Today's search for oil and gas has taken the drilling industry into the world's most remote areas. Wells have been drilled under a wide range of environmental conditions, including the most hostile in the world.

In addition to drilling in water depths of several thousand feet, offshore wells have been drilled in seas where maximum wave heights can reach 90 ft and where storms are among the most severe ever encountered. Offshore wells have been drilled in Arctic waters from man-made islands (Fig. 1-4) by deviating the well from the vertical under controlled conditions. In some areas, wells have been drilled from floating drilling units while constant watch was maintained for moving icebergs. Plans for abandoning the well or taking other steps as an iceberg nears the rig are laid out before the rig is ever placed on location.

On land, wells are drilled in the world's large deserts where no roads exist. The rig must be moved to location on special carriers. Supply and crew transportation is often only by air. Rigs have also been specially designed to be carried in modules by helicopters and landed in jungle areas. In some cases, bulldozers must be disassembled, flown by helicopter to these locations, and reassembled before a drilling site can be cleared. Other difficult conditions for drilling are caused by natural or man-made obstacles, and it is necessary to drill directionally—slant the hole—because the rig cannot be erected where desired.

All of this experience has shown that a well can be drilled virtually anywhere a commercial deposit of oil or gas is suspected. It may be difficult and enormously expensive, but the drilling industry has the know-how and the personnel to do it.

Types of wells

Although the equipment used to drill most oil and gas wells is quite similar, there are a number of different types of wells, depending on their purpose and their relationship to other wells and fields.

A wildcat well is a well drilled in an unproven area, far from any existing producing well.[11] It is an exploratory well in the truest sense of

Fig. 1–4 Rig drilling on gravel island in Arctic waters.

the word, although the term *exploratory well* may also refer to a well that is not as remote from existing proven reserves as a wildcat is. An exploratory well may also be one that is drilled to find the limits of an oil- or gas-bearing formation or pool.

A stepout well is a well drilled adjacent to a proven well but located in an unproven area. It is also drilled in an attempt to determine the boundaries of a producing formation. Development wells, on the other hand, are drilled in an area already proven to be productive. Development wells are drilled after the discovery well is drilled to exploit an oil or gas field. The discovery well is an exploratory well that encounters a new or previously untapped oil or gas deposit; it is a successful wildcat.

Infill wells are wells drilled to fill in between established producing wells on a lease. They reduce the spacing between wells in order to increase production from the lease.

Other common industry terms are also of interest. A dry hole is a well in which no significant amount of oil or gas is found. In some statistical reviews and surveys, however, offshore discovery wells drilled with a mobile drilling unit may be listed as dry when they did encounter oil or gas.[12] It is often more economical to plug such a well and drill development wells from a fixed platform. Therefore, these discovery wells may be counted as dry because they were plugged and abandoned. Only a handful of such wells is included in the thousands drilled each year.

An oil well produces hydrocarbons existing in the underground reservoir in a liquid form; a gas well produces hydrocarbons existing initially in a gaseous phase in the reservoir.[13] Production from an oil well will contain hydrocarbons in the gaseous phase, and production from a gas well will contain liquids. The ratio of gas to liquid varies widely from field to field and may change significantly in a single well over the well's life. A gas-condensate well, though normally classed as a gas well, produces large amounts of hydrocarbon liquids along with the gas.

A stripper well is normally a well in its later stage of production and is defined—for various purposes, including pricing and taxation—as an oil well producing 10 b/d or less. In 1976, there were nearly 400,000 stripper wells in the U.S.

These classifications have little to do with how an oil or gas well is drilled, only with where it is drilled and for what purpose. A deep exploratory well requires a larger rig than a shallow development well, and the large rig may have special auxiliary equipment because drilling conditions are uncertain. Additional equipment may be needed if a well will be drilled directionally under some surface obstruction rather than vertically.

The parties involved

In most cases, the owner of the drilling rig has no equity interest in the well but serves as a contractor to the well owners. There are exceptions; some contractors own equity in a few of the wells they drill, but this is not common.

For much of the industry's history, the well owners—primarily the oil-producing companies—owned their own drilling rigs and drilled their own wells. But few producing companies own rigs now. In the 1950s and 1960s they decided that having an outside firm whose sole business was drilling do the work was a more cost-effective approach.

When the well is owned by only one firm, that firm is referred to as the operator. When drilling is complete and the well is put on production, that company will operate the well. The drilling contractor, having fulfilled his contract obligation to drill the well, will move to another drilling location.

There are often several owners in a single oil or gas well. In this case, one of the owners (firms) will be designated the operator by agreement of all of the owners. Often the operator is the owner with the largest equity ownership in the well.

Before drilling can begin, the well owners must obtain an oil and gas lease from the owners of the minerals. This transaction can involve a variety of financial agreements. Typically, though, the minerals owner receives a cash bonus from the well owners and a royalty interest in production from the well to be drilled. The minerals owner's royalty is a share of the gross production of oil and gas from wells on his property before deducting the cost of producing the oil or gas. In the past, this royalty has typically been one-eighth of the gross production. As drilling activity grew rapidly in recent years and the demand for acreage on which to drill increased, higher royalties have been negotiated.

Sometimes the owner of the minerals is not the owner of the land's surface. In this case, the well owners (operator) must still obtain a lease from the minerals owner but in addition must negotiate with the surface owner for permission to move in drilling equipment and drill the well.

In addition to the lease, permits may be required from government authorities. These permits may be needed for disposal of drilling fluids, for drilling on federal lands, or to meet other regulations.

When all permits are obtained, the contractor can move the rig on location and begin drilling.

Obtaining offshore acreage on which to drill is somewhat different than leasing onshore acreage from a private party. Most offshore areas are administered by the government of the nation bordered by the body of water. Large blocks of offshore acreage are usually involved because of the expense of exploring and developing offshore fields. On land, a lease may include a relatively small area.

The agreements reached in acquiring offshore blocks for oil and gas exploration vary widely, depending in part on the national goals of the nation involved. In the U.S., most of the acreage made available is leased to the firm offering the highest cash bonus during a competitive lease sale. In addition, the successful bidder pays the federal government a royalty on all oil and gas produced from the block. Other systems have been used, including a variable-royalty bidding system. Some countries may impose different requirements on companies desiring to explore in their waters.

The amounts paid in bonuses in competitive bidding in the U.S. can be large, depending on whether the acreage being offered is proven, how the bidders view the chances of finding oil or gas, and general industry economic conditions. For instance, a recent federal offering of tracts in the central and western Gulf of Mexico netted cash bonuses of $2.633 billion.[14] In the sale, 87 companies bid on 162 tracts of the 212 tracts offered for lease. Tracts totaling more than 830,000 acres received bids. This is one of the largest cash bonus totals offered in a single U.S. sale, but over the years many billions of dollars have been paid in bonus money for offshore leases.

The federal government also owns land in the U.S. A significant number of onshore wells in the West are drilled on these government-owned lands. Some of this land is offered for lease through competitive bidding, mainly where proven reserves exist. And some, in unproven areas, is offered for lease under a noncompetitive system known as *simultaneous leasing*. In this system, certain tracts are made available several times each year. An application to lease any of the tracts offered can be filed by any interested party for a $75 filing fee. When the filing period (usually about three weeks) is closed, the winner of each lease is determined by conducting a drawing, or lottery. Those receiving leases in the drawing must pay an annual rental fee on the lease.

Many individuals and companies besides oil producers participate in the simultaneous leasing program. If successful, they attempt to sell their lease for cash, an overriding royalty, or both to someone who wants to drill a well on the lease. Of course, the federal government also receives a royalty on all oil and gas production from the lease.

Supply and service firms. Throughout the drilling operation, the services of a wide variety of firms are necessary to complete the well.

And equipment that will remain in the well—casing, tubing, pumps, and wellhead—is provided by oilfield equipment supply companies.

Specialized services that the drilling contractor does not provide but which may be necessary to drill and complete the well may include:

1. casing installation crews and equipment
2. casing cementing
3. directional drilling services and equipment
4. fishing services and tools
5. specialty tools such as downhole motors
6. inspection services
7. logging and other analysis services and equipment
8. trucking

This incomplete list indicates the large number of separate firms and specialized services that may be required to drill a single oil or gas well.

The drilling crew. The rig crew consists of a driller, floormen, and a derrickman. The driller is in charge of the rig crew and operates the rig controls. Duties of the floormen include connecting joints of drill pipe while drilling—making a connection—and unscrewing and screwing together the pipe sections when the drill string is being removed from the hole or lowered into the hole. They also perform other functions, depending on the specific drilling operation being conducted.

The derrickman's primary job is to handle the top end of the sections of pipe being pulled from the hole during a trip into or out of the hole. He works from a platform high in the derrick during this operation. He also performs other duties around the rig site, as do other members of the rig crew.

Each crew works a shift, usually eight hours on a conventional land rig. Shift lengths vary, however, on offshore rigs and in special remote locations.

A toolpusher is the rig's supervisor and has charge of all drilling crews working on the rig. He may be in charge of more than one rig if the rigs are of modest size and the holes they are drilling are routine. In some cases, though, such as on a large rig drilling a difficult well, one toolpusher may be assigned only one rig.

In addition to these key personnel who are all on the drilling contractor's payroll, the operator or well owner has representatives at the well site, including engineers and geologists. At any given time, other firms that are providing special services or equipment will have representatives at the well to operate and maintain their equipment.

16 FUNDAMENTALS OF DRILLING

On mobile offshore drilling units, the staff is considerably larger. Since the mobile drilling unit is a vessel, it must have its own crew in addition to the drilling crews. And because crews are quartered on offshore drilling units, cooks and galley personnel are needed. Electricians, mechanics, and other specialized personnel may be aboard, in addition to the company representatives already mentioned.

It is impossible to list all of the personnel who play a part in drilling a single well. Nevertheless, these are key people directly involved in the drilling operation.

Payment basis. The drilling contractor is usually paid based on either the number of days worked (a day rate) or the number of feet drilled (footage rate). He bids on the job on one or the other of these bases.

On a day-rate basis, the drilling contractor is paid by the day for the number of days required to drill the well. Under a footage contract, the drilling contractor is paid an agreed-upon amount per foot of hole drilled. The price per foot may vary for different sections of the hole, and some contracts call for a combination of day rates and footage rates in a single well. Such a combination contract might call for a certain footage rate until the well reached a specific depth; then the drilling contractor would be paid on a day-rate basis from that point to total depth.

This is not uncommon in deep wells and in wells where drilling conditions are either uncertain or difficult in the deeper zones. It may also be done where the operator wants to be free to test some zones extensively and is uncertain how long that testing will take. By paying the contractor on a day-rate basis in this interval, the operator can take whatever time is warranted for testing or other operations without loss of revenue or profit by the contractor.

Some wells are drilled under a turnkey contract, in which the drilling contractor agrees to furnish all materials and labor and to do all that is required to drill and complete the well. This type of contract is used primarily for wells where drilling conditions can be accurately predicted and where the contractor is confident in his ability to estimate his cost and the time required to drill the well.

Key industry statistics

The single most important factor influencing the number of wells that are drilled is the price of oil and gas. The number of wells drilled each year has fluctuated widely, especially in the U.S.; but some trends are apparent.

U.S. oil and gas well completions (wells drilled) reached the first peak in 1956 when 58,160 wells were completed (Table 1–1). Following that record year, the number of wells drilled began a long decline. The low point in that decline occurred in 1971 when only 27,300 oil and gas wells were completed in the U.S. At that time, the wellhead price of oil was about $3.50/bbl, and the average price of natural gas at the wellhead in the U.S. was less than 20¢/Mcf (thousand cubic feet).

In 1973, the Arab embargo and subsequent price increases sparked a modern-day resurgence in drilling activity. The number of wells drilled each year began to climb steadily. Further price increases in 1978

TABLE 1–1
Quarter-century historical record of U.S. well completions

Year	Total wells	Total footage	Total wildcats
1981*	80,031	362,538,000	16,565
1980	62,462	288,936,000	11,916
1979	51,263	243,207,903	10,484
1978	48,513	231,383,981	10,677
1977	46,479	215,010,591	9,961
1976	41,455	185,344,695	9,234
1975	39,097	178,505,570	9,214
1974	32,893	153,164,191	8,619
1973	27,602	138,937,944	7,466
1972	28,755	138,285,876	7,539
1971	27,300	128,335,404	6,922
1970	29,467	142,431,468	7,693
1969	34,053	160,949,360	9,701
1968	32,914	149,287,860	8,879
1967	33,558	144,234,721	8,886
1966	37,881	166,025,335	10,313
1965	41,423	181,427,015	8,265
1964	45,236	189,921,870	9,258
1963	43,653	184,357,230	8,607
1962	46,179	198,558,641	9,003
1961	46,962	192,116,114	9,191
1960	46,751	190,702,672	9,635
1959	51,764	209,231,416	10,073
1958	50,039	198,224,092	9,588
1957	55,024	223,087,055	11,739
1956	58,160	233,902,000	13,034
1955	56,682	226,270,000	12,271
1954	56,930	218,986,112	11,280

*Estimated
Source: *Oil & Gas Journal*, 25 January 1982, p. 147

18 FUNDAMENTALS OF DRILLING

moved drilling activity into the boom category. The number of wells drilled in the U.S. continued to climb, and an estimated 80,031 oil and gas wells were completed in 1981 (Table 1-2). That far surpassed the mid-1950s peak.

A world price for oil of around $35/bbl following decontrol of oil prices in the U.S. was responsible for the record activity. Although prices were stabilized in early 1982 at that level, the outlook was for a continued high level of drilling activity. Along with the dramatic rise in oil prices, natural gas prices also increased in a few short years. Despite a complex structure of regulated gas prices, drilling for both gas and oil continues at a high level. One forecast put the number of well completions in the U.S. at more than 89,000 in 1982.[15]

Directly related to the number of wells drilled is the number of rotary

TABLE 1-2
Well forecast for 1982

	1981 estimate			1982 Forecast			
State	Total wells 1981	Total wildcats 1981	Total footage (1,000 ft)	Total wells 1982	Total wildcats 1982	Field wells 1982	Total footage (1,000 ft)
Alabama	214	75	1,187	250	100	150	1,386
Alaska	155	25	1,551	246	56	190	2,461
Arizona	7	6	36	10	10	0	51
Arkansas	905	200	4,463	960	220	740	4,735
Atlantic, Offshore	12	12	204	12	11	1	204
California	3,227	370	8,755	3,950	462	3,488	11,034
Onshore	2,960	350	7,400	3,500	412	3,088	8,750
Offshore	267	20	1,355	450	50	400	2,284
Colorado	1,740	766	9,645	1,900	815	1,085	10,532
Florida	25	11	295	39	20	19	461
Georgia	3	3	21	1	1	0	7
Idaho	5	5	45	8	8	0	73
Illinois	3,074	626	8,374	4,000	690	3,310	10,896
Indiana	850	345	1,601	990	360	630	1,864
Kansas	6,850	2,100	22,810	7,671	2,280	5,391	25,544
Kentucky	1,900	550	2,538	2,060	610	1,450	2,752
Louisiana	5,800	1,309	37,269	6,340	1,505	4,835	41,268
North	3,600	485	13,082	3,860	560	3,300	14,027
South	1,500	625	17,021	1,670	725	945	18,949
Offshore	700	199	7,166	810	220	590	8,292
Maryland	1	1	-	1	1	0	-
Michigan	730	307	2,925	790	320	470	3,166
Mississippi	785	360	7,086	849	390	459	7,664
Missouri	190	32	63	165	20	145	55
Montana	1,050	425	4,952	1,100	495	605	5,188

TABLE 1-2 Continued

State	1981 estimate			1982 Forecast			
	Total wells 1981	Total wildcats 1981	Total footage (1,000 ft)	Total wells 1982	Total wildcats 1982	Field wells 1982	Total footage (1,000 ft)
Nebraska	580	250	2,890	630	290	340	3,139
Nevada	26	22	163	43	20	23	270
New Mexico	2,590	430	14,709	2,960	520	2,440	16,834
East	1,300	250	7,738	1,550	300	1,250	9,226
West	1,290	180	6,971	1,410	220	1,190	7,608
New York	600	115	1,745	600	120	480	1,750
North Dakota	735	370	6,574	845	425	420	7,558
Ohio	4,581	71	15,400	4,700	90	4,610	14,900
Oklahoma	10,420	650	44,806	13,185	900	12,285	54,058
Oregon	14	7	50	14	7	7	50
Pennsylvania	3,500	380	8,757	3,680	420	3,260	9,207
South Dakota	80	48	294	90	80	10	331
Tennessee	350	190	595	350	200	150	595
Texas	24,966	5,572	130,263	26,655	6,110	20,545	141,340
District 1	2,451	323	9,522	2,500	380	2,120	9,713
District 2	865	412	5,860	1,000	430	570	6,775
District 3	2,560	680	21,102	2,700	710	1,990	22,259
District 4	1,400	570	10,228	1,565	610	995	11,434
District 5	800	170	5,166	825	185	640	5,328
District 6	985	234	7,488	1,020	265	755	7,754
District 7-B	4,400	1,319	14,023	4,600	1,400	3,200	15,212
District 7-C	1,965	446	10,772	2,160	500	1,660	11,841
District 8	2,411	290	13,282	2,680	340	2,340	15,745
District 8-A	1,945	372	11,312	2,100	420	1,680	12,214
District 9	3,718	575	12,106	3,900	600	3,300	12,698
District 10	1,230	70	7,208	1,325	140	1,185	7,764
Offshore	236	111	2,194	280	130	150	2,603
Utah	445	186	2,726	509	220	289	3,119
Virginia	25	1	102	30	4	26	122
Washington	0	0	0	6	6	0	-
West Virginia	1,911	165	7,644	2,095	195	1,900	8,380
Wyoming	1,685	580	12,000	1,900	650	1,250	13,505
Total U.S.	80,031	16,565	362,538	89,634	18,631	71,003	404,499
Canada	6,664	3,148	25,666	7,000	3,146	3,854	26,959
Western Canada	6,652	3,136	25,499	6,990	3,138	3,852	26,819
Alberta	5,600	2,400	21,521	5,900	2,385	3,515	22,680
British Columbia	200	150	1,470	210	160	50	1,543
Saskatchewan	790	550	2,260	820	555	265	2,350
N.W.T.	9	7	75	10	8	2	82
Manitoba	53	29	173	50	30	20	164
East Coast Offshore	12	10	167	10	8	2	140

Source: *Oil & Gas Journal*, 25 January 1982, p. 146

rigs operating. The drilling boom of the late 1970s and early 1980s brought a corresponding increase in the number of rigs in the industry's fleet. At the bottom of the slump in 1971, the average number of rotary rigs operating in the U.S. was less than 1,000 (Table 1–3). About half that many were operating in other countries around the world.[16] As oil and gas began commanding higher prices, operators wanted to drill more wells. To meet the demand for more rigs, a rig-building scramble began.

In 1981, the average number of operating rotary rigs in the U.S. stood at almost 4,000 and was expected to increase even more in 1982. As operator demands for more rigs grew, the number of firms offering rig contracting services also grew. In late 1981, the International Association of Drilling Contractors (IADC) had 1,043 contractor members, up from 732 a year earlier.

Well costs. As the price of oil and gas increased, so did the cost of drilling. Individual well drilling costs vary widely. At the least expensive end of the cost range is a shallow well drilled on land in a proven field in an easily accessible location. At the extremely high end of the spectrum is a deep exploratory well drilled in deep water or in a remote area—in the Arctic, for example.

The almost infinite variety of conditions must be kept in mind when

TABLE 1–3
Yearly average active rotary rigs in the U.S.

Year	Average number active rigs	Year	Average number active rigs
1951	2,543	1967	1,134
1952	2,641	1968	1,150
1953	2,613	1969	1,194
1954	2,509	1970	1,028
1955	2,687	1971	975
1956	2,619	1972	1,095
1957	2,429	1973	1,194
1958	1,923	1974	1,471
1959	2,074	1975	1,660
1960	1,750	1976	1,658
1961	1,760	1977	2,001
1962	1,636	1978	2,258
1963	1,501	1979	2,177
1964	1,502	1980	2,910
1965	1,387	1981	3,969
1966	1,273		

Source: *Oil & Gas Journal,* 28 January 1963, 29 January 1973, and 25 January 1982

TABLE 1-4
International active rotary drilling rigs

Region	December 1981			December 1980		
	Land	Offshore	Total	Land	Offshore	Total
NORTH AMERICA*						
Canada	262	5	267	339	4	343
U.S.	4,259	271	4,530	3,077	249	3,326
Subtotal	4,521	276	4,797	3,416	253	3,669
LATIN AMERICA						
Argentina	58	4	62	80	2	82
Barbados	1	0	1	0	0	0
Bolivia	14	0	14	4	0	4
Brazil	60	29	89	43	32	75
Chile	3	3	6	4	4	8
Colombia	23	0	23	20	1	21
Guatemala	7	0	7	6	0	6
Dominican Republic	1	0	1	0	0	0
Ecuador	5	0	5	3	0	3
Mexico	201	19	220	194	22	216
Paraguay	1	0	1	0	0	0
Peru	21	5	26	17	5	22
Trinidad	8	6	14	7	5	12
Venezuela	51	16	67	36	14	50
Subtotal	454	82	536	414	85	499
EUROPE						
Austria	10	0	10	7	0	7
Denmark	0	2	2	2	0	2
France	28	2	30	16	2	18
Germany	30	0	30	21	1	22
Greece	3	3	6	1	2	3
Holland	9	9	18	5	9	14
Italy	25	8	33	17	8	25
Norway	0	12	12	0	11	11
Portugal	1	0	1	0	0	0
Spain	9	3	12	7	5	12
Switzerland	1	0	1	1	0	1
U.K.	8	55	63	2	48	50
Yugoslavia	22	0	22	22	0	22
Subtotal	146	94	240	101	86	187
AFRICA						
Algeria	77	0	77	106	0	106
Angola	2	6	8	0	4	4
Cameroon	0	6	6	0	8	8
Chad	1	0	1	1	0	1
Congo	2	5	7	1	4	5
Gabon	4	11	15	2	4	6

22 FUNDAMENTALS OF DRILLING

TABLE 1-4 Continued

Region	December 1981			December 1980		
	Land	Offshore	Total	Land	Offshore	Total
Ivory Coast	0	6	6	1	2	3
Libya	31	2	33	38	2	40
Morocco	5	1	6	2	1	3
Niger	5	0	5	5	0	5
Nigeria	19	9	28	15	5	20
South Africa	0	2	2	0	2	2
Egypt	11	19	30	11	16	27
Sudan	2	0	2	2	0	2
Tunisia	8	3	11	7	2	9
Zaire	2	1	3	1	0	1
Subtotal	169	71	240	192	50	242
MIDDLE EAST						
Abu Dhabi	19	16	35	5	19	24
Bahrain	1	0	1	1	0	1
Dubai	1	2	3	1	2	3
Iraq	5	0	5	39	0	39
Kuwait	3	0	3	3	0	3
Oman	11	0	11	8	0	8
Qatar	3	3	6	4	4	8
Ra's al Khaymah	1	0	1	1	0	1
Saudi Arabia	16	13	29	15	10	25
Sharjah	2	1	3	1	0	1
Syria	16	0	16	15	0	15
Turkey	26	0	26	40	0	40
Umm al Qaiwain	1	0	1	0	0	0
Subtotal	105	35	140	133	35	168
FAR EAST						
Bangladesh	4	0	4	5	0	5
Brunei	1	6	7	1	6	7
Burma	24	0	24	28	0	28
China	0	4	4	0	13	13
India	37	7	44	38	5	43
Indonesia	60	29	89	61	23	84
Japan	15	2	17	13	2	15
Malaysia	0	13	13	0	13	13
Pakistan	14	0	14	15	0	15
Philippines	14	3	17	8	2	10
Sri Lanka	0	1	1	0	0	0
Taiwan	11	1	12	11	1	12
Thailand	3	4	7	1	4	5
Subtotal	183	70	253	181	69	250

TABLE 1-4 Continued

Region	December 1981			December 1980		
	Land	Offshore	Total	Land	Offshore	Total
SOUTH PACIFIC						
Australia	29	9	38	11	3	14
New Zealand	1	1	2	3	1	4
Papua New Guinea	1	0	1	0	0	0
Subtotal	31	10	41	14	4	18
Total	5,609	638	6,247	4,451	582	5,033

*Rig count as of December 28, 1981, and December 29, 1980
Source: Hughes Tool Co. and *Oil & Gas Journal*, 25 January 1982, p. 114.

considering an average well cost. In 1979, the average drilling cost for all wells drilled in the U.S. was $331,367 per well, or almost $68/ft of hole drilled.[17] The cost of drilling oil wells averaged $58.29/ft and the cost of drilling a gas well averaged $80.66/ft, reflecting the fact that gas wells generally are deeper than oil wells. Tables 1-5 and 1-6 show how widely costs vary from these averages.

Offshore wells are much more costly than those drilled on land. In 1979, for instance, the average offshore well cost about $2.5 million to drill and equip. Average depth was 9,835 ft. Some of the highest-cost wells were drilled offshore Alaska that year, where three wells averaged $8.09 million each. Off the Atlantic coast, where 18 dry holes were drilled, the average cost was $7.7 million/well.

Wells drilled outside the U.S., as a rule, are more costly. Many are remote from established service and supply points, and transportation expenses, for example, can be much higher.

The average depth of the wells drilled in a given year must also be considered in light of the wide range of individual well depths. In 1979, the average depth of all wells drilled in the U.S. was 4,894 ft, but individual wells ranged in depth from a few hundred feet to more than 20,000 ft.

As might be expected, average well depth does not change significantly from year to year because the largest number of wells drilled are of relatively modest depth. But changes in average well depth can be one indicator of whether the industry is concentrating on deep exploration drilling or further development of existing shallower fields. Other indicators of industry trends are the ratio of exploratory wells to development wells and the ratio of oil wells drilled to gas wells drilled. If oil

TABLE 1-5
Estimated cost of drilling and equipping onshore wells, 1980
(average cost/well in $1,000)

	Depth interval, ft	Oil	Gas	Dry	Total
No. wells		4,827	654	1,736	7,217
Ave. depth	0–1,249	851	935	7474	833
Ave. cost/well		31.6	33.3	21.9	29.4
No. wells		5,480	2,485	2,608	10,573
Ave. depth	1,250–2,499	1,829	2,027	1,882	1,889
Ave. cost/well		73.4	70.3	52.9	67.6
No. wells		5,777	2,546	3,507	11,830
Ave. depth	2,500–3,749	3,127	3,215	3,161	3,156
Ave. cost/well		128.3	128.1	81.1	114.3
No. wells		3,775	2,581	3,091	9,447
Ave. depth	3,750–4,999	4,397	4,293	4,334	4,348
Ave. cost/well		190.9	164.0	118.4	159.8
No. wells		4,039	2,959	3,348	10,346
Ave. depth	5,000–7,499	6,099	6,074	6,091	6,089
Ave. cost/well		324.8	303.9	217.0	284.0
No. wells		3,280	1,663	1,736	6,679
Ave. depth	7,500–9,999	8,643	8,613	8,620	8,630
Ave. cost/well		617.0	631.0	479.0	584.6
No. wells		764	1,077	936	2,777
Ave. depth	10,000–12,499	10,947	11,022	11,118	11,034
Ave. cost/well		1,204.3	1,192.2	947.5	1,113.0
No. wells		241	473	445	1,159
Ave. depth	12,500–14,999	13,308	13,539	13,536	13,490
Ave. cost/well		1,833.7	2,120.6	1,737.1	1,913.7
No. wells		45	233	211	489
Ave. depth	15,000–17,499	15,997	16,090	16,156	16,110
Ave. cost/well		3,363.8	3,450	3,215.4	3,341.1
No. wells		8	65	71	144
Ave. depth	17,500–19,999	18,317	18,533	18,392	18,452
Ave. cost/well		7,319.1	5,511.0	4,329.6	5,028.9
No wells		—	45	33	78
Ave. depth	20,000 and over	—	21,154	21,285	21,209
Ave. cost/well		—	8,381.2	8,447.2	8,409.1
Total no. wells		28,236	14,781	17,722	60,739
Ave. depth all wells		4,405	5,507	4,959	4,667
Ave. cost/well all wells		245.2	454.7	299.6	312.1

Source: 1980 Joint Association Survey on Drilling Costs, December 1981 edition

THE OIL AND GAS WELL DRILLING INDUSTRY 25

TABLE 1-6
Estimated cost of drilling and equipping offshore wells, 1980
(average cost/well in $1,000)

	Depth interval, ft	Oil	Gas	Dry	Total
No. wells		—	—	3	3
Ave. depth	0–1,249	—	—	1,055	1,055
Ave. cost/well		—	—	714.1	714.1
No. wells		—	2	4	6
Ave. depth	1,250–2,499	—	2,282	2,134	2,183
Ave. cost/well		—	1058.4	842.6	914.5
No. wells		11	7	9	27
Ave. depth	2,500–3,749	3,157	3,390	3,183	3,226
Ave. cost/well		618.2	1,496.7	967.3	962.4
No. wells		36	15	14	65
Ave. depth	3,750–4,999	4,377	4,300	4,314	4,346
Ave. cost/well		771.4	2,307.5	1,135.3	1,204.2
No. wells		63	73	78	214
Ave. depth	5,000–7,499	6,346	6,303	6,491	6,384
Ave. cost/well		1,514.9	2,007.0	1,464.6	1,664.5
No. wells		94	121	125	340
Ave. depth	7,500–9,999	8,689	8,871	8,732	8,769
Ave. cost/well		2,268.5	2,739.9	1,786.8	2,259.9
No. wells		61	139	152	352
Ave. depth	10,000–12,499	11,315	11,138	11,136	11,168
Ave. cost/well		3,647.2	3,259.9	2,963.8	3,199.1
No. wells		38	61	77	176
Ave. depth	12,500–14,999	13,533	13,543	13,585	13,559
Ave. cost/well		4,801.2	4,794.1	4,930.6	4,855.4
No. wells		13	24	41	78
Ave. depth	15,000–17,499	15,895	15,854	16,052	15,965
Ave. cost/well		6,759.3	6,657.5	6,896.7	6,800.2
No. wells		1	2	6	9
Ave. depth	17,500–19,999	17,905	18,732	18,331	18,373
Ave. cost/well		8,262.1	7,571.6	9,763.7	9,109.7
No wells		—	—	2	2
Ave. depth	20,000 and over	—	—	20,449	20,449
Ave. cost/well		—	—	8,404.2	8,404.2
Total no. wells		317	444	511	1,272
Ave. depth all wells		8,952	9,952	10,266	9,829
Ave. cost/well all wells		2,663.4	3,256.2	3,045.5	3,023.8

Source: 1980 Joint Association Survey on Drilling Costs, December 1981 edition

prices—on an equivalent basis related to heating value—surge ahead of gas prices, operators will tend to put more money into drilling where they expect to find oil. When the relationship shifts back in favor of gas, more wells may be drilled in search of gas.

Deepest wells. The deepest well in the U.S., drilled to a total depth of 31,441 ft, is located in Oklahoma. Completed in 1974, it was the second well drilled in the U.S. to a depth of more than 30,000 ft.

In Russia, a well is underway that is planned to reach a total depth of 15,000 meters (49,212 ft). The well has been in progress since 1970; in November 1981 it had passed a depth of 36,000 ft.

These few ultra-deep wells are not necessarily the beginning of a trend to extreme-depth wells. Some experts believe the chances of commercial hydrocarbons existing at these depths are slim. And there are many areas of the world yet to be explored that may contain oil and gas reserves at much more modest depths. Still, these very deep wells indicate what the industry can do. Experience gained at these depths has been valuable in designing tools for special conditions in other areas at lesser depths and in planning other deep wells.

Extreme temperatures and pressures, the weight of drill pipe needed to reach to 30,000 ft or more, and the extreme weight of casing for one of these wells are key technical constraints on drilling beyond today's record depths. There is no doubt that the necessary equipment could be built and a well could be drilled to even greater depths. The cost and the prospect of finding oil or gas will be the factors that determine whether or not it is done.

The only real test. Techniques used to evaluate areas that may contain oil or gas have developed rapidly, especially since the availability of computer analysis techniques.

Geophysical methods are infinitely more sophisticated and more accurate than only a few years ago. These and other techniques can provide much information about underground reservoirs by taking measurements at the surface. But the only way to be certain that oil or gas is present—and in what quantity—is to drill a well.

Evidence that exploration methods other than actual drilling are not foolproof is the number of nonproductive holes still drilled each year. These dry holes do not reflect badly on the capability of the drilling industry or on the talents of explorationists. Rather, they confirm that the only way to find oil or gas is to drill a well.

TABLE 1-7
How U.S. drilling and completion costs have risen*

	Average index for the year (1979 = 100)															Prelim 1980	Percent Increase 1970-80
	1965	1966	1967	1968	1969	1970	1971	1972	1973	1974	1975	1976	1977	1978	1979		
Payments to drilling contractors	20.2	21.1	21.5	23.6	25.6	26.7	27.7	31.2	34.0	44.2	50.7	57.9	69.4	78.5	100.0	108.0	304.4
Purchased items																	
Road and site preparation†	38.2	39.2	42.0	43.9	46.4	49.1	52.7	56.3	59.8	64.1	71.1	76.5	82.5	90.1	100.0	111.7	127.5
Transportation†	41.7	42.8	45.8	53.2	53.7	56.0	59.6	61.2	63.0	70.0	76.1	80.8	86.5	93.8	100.0	103.8	85.4
Fuel*	18.5	19.3	20.5	19.6	19.8	21.3	22.0	22.3	25.4	44.5	54.1	62.9	76.2	81.7	100.0	146.3	586.9
Drilling mud and additives†	27.9	28.1	29.8	31.0	32.2	35.1	37.9	38.0	42.6	50.5	64.5	72.5	76.4	90.6	100.0	117.1	233.6
Well site logging and or monitoring system	39.2	41.1	41.4	42.8	47.1	47.9	48.4	48.4	48.4	55.2	64.9	69.6	75.2	85.3	100.0	112.0	133.8
All other physical tests	32.2	33.5	35.0	36.8	39.4	41.9	45.3	45.3	47.1	53.4	64.2	72.3	79.3	87.2	100.0	112.7	169.0
Logs and wire line evaluation services†	26.0	27.1	29.4	31.3	33.3	36.1	39.3	39.3	40.6	45.2	53.4	62.4	69.0	79.2	100.0	109.8	204.2
Directional drilling services	44.2	44.2	44.2	44.2	44.9	47.0	51.1	52.2	54.2	61.8	65.9	71.9	78.0	87.3	100.0	114.3	143.2
Perforating	39.2	39.3	42.6	43.0	45.6	49.8	51.6	51.7	51.7	57.6	68.1	75.5	82.4	89.4	100.0	120.4	141.8
Formation treating†	40.6	43.8	45.5	47.1	49.3	50.7	54.0	54.2	55.8	59.7	75.6	82.0	86.0	92.1	100.0	120.8	138.3
Cement and cementing services	35.6	37.6	39.7	42.2	44.0	46.9	52.4	53.3	55.8	60.6	75.5	81.0	83.1	92.3	100.0	116.8	149.0
Casing and tubing	27.7	27.7	27.7	30.6	33.3	35.2	41.8	47.4	47.3	64.2	71.4	77.3	85.2	94.8	100.0	114.8	226.1
Casing hardware	27.7	27.7	30.6	33.3	33.4	35.2	41.8	47.4	47.3	64.2	71.4	77.3	85.2	94.8	100.0	114.8	226.1
Special tool rentals	41.8	42.1	43.2	44.6	46.4	50.9	52.2	52.2	53.0	59.0	67.8	75.0	82.1	90.3	100.0	108.2	112.6
Drill bits and reamers	32.7	33.0	34.3	37.7	40.8	43.9	44.5	47.1	48.1	54.9	68.2	73.7	81.2	90.9	100.0	114.1	159.9
Wellhead equipment	27.9	29.0	30.0	32.2	33.8	35.5	37.2	37.8	40.0	46.7	56.3	66.1	77.2	86.1	100.0	120.2	238.6
Other equipment and supplies	32.9	33.4	34.5	36.7	38.9	40.8	42.3	44.0	46.0	54.5	67.8	75.1	81.6	90.3	100.0	115.7	183.6
Plugging	34.6	36.5	38.7	41.4	44.5	49.0	55.1	59.0	61.5	66.1	76.0	80.6	84.6	92.9	100.0	110.8	126.1
Supervision and overhead	35.5	37.1	39.3	41.6	44.3	46.2	48.6	51.4	54.8	62.4	69.2	74.5	80.7	90.8	100.0	110.7	139.6
All other expenditures	40.8	41.6	42.3	43.3	44.8	46.5	48.2	49.9	53.3	65.0	72.5	77.0	82.5	88.6	100.0	115.9	149.2

*Unadjusted for depth, 1979 = 100. Some figures are approximate due to change in base year from 1974 to 1979. †Subject to further revision due to revised weightings within these items.

Source: Independent Petroleum Association of America based on data from various sources and *Oil & Gas Journal*, 26 October 1981, p. 64.

References

1. Moore, W.D., III. "Ingenuity Sparks Drilling History." Petroleum 2000. *Oil & Gas Journal*. (August 1977), p. 159.
2. Brantley, J.E. *History of Oil Well Drilling*. Houston: Gulf Publishing Co., 1971.
3. Maurer, William C. *Advanced Drilling Techniques*. Tulsa: PennWell Publishing Co., 1980.
4. Vielvoye, Roger, and Richard Wheatley. "Oil Prices Hold Key to Growth of Industry Action in Deep Water." *Oil & Gas Journal*. (11 January 1982), p. 25.
5. "Deep Ocean Reentry a Scientific First." *Oil & Gas Journal*. (18 January 1971), p. 28.
6. Burnett, Barbara. "Subsea Completions Should Number 140 by Year End." *Offshore*. (August 1978), p. 47.
7. "Global Action Points to Gain in Offshore Production." *Oil & Gas Journal*. (9 February 1981), p. 27.
8. Moore, W.D., III. "Offshore Drilling Maintains Fast Pace." *Oil & Gas Journal*. (3 May 1982), p. 143.
9. "Contractors Place Orders for 19 New Platform Rigs." *Offshore*. (January 1982), p. 87.
10. 1982 Market Data Guide. *Oil & Gas Journal*. (August 1981).
11. Langenkamp, Robert D. *Handbook of Oil Industry Terms and Phrases, 3rd Edition*. Tulsa: PennWell Publishing Co., 1981.
12. *Quarterly Review of Drilling Statistics for the United States*. American Petroleum Institute (February 1981).
13. *1977 Joint Association Survey On Drilling Costs*. American Petroleum Institute. (February 1979).
14. "Gulf of Mexico Sale Nearly Sets Record." *Oil & Gas Journal*. (27 July 1981), p. 103.
15. McCaslin, John C. "1982: Biggest Drilling Year in U.S. History." *Oil & Gas Journal*. (25 January 1982), p. 145.
16. Lange, David. "1978: Oil's Biggest Volume Year Yet." *Oil & Gas Journal*. (30 January 1978), p. 119.
17. "U.S. Drilling Outlay Jumps 23% During 1979." *Oil & Gas Journal*. (16 March 1981), p. 43.

2 Types of Rigs

THE demands of the drilling industry for rigs that can operate in almost any environment and drill to depths in excess of 30,000 ft has resulted in a number of different types of designs. Although most rigs use the rotary method of drilling, they are available in a variety of sizes for different depth requirements, in several designs to fit transportation needs, and with a range of equipment for special drilling jobs. Of course, offshore rigs must be supported by a fixed or mobile platform while drilling, and several broad types of these platforms are used.

One key reason for the range in depth capability is the fact that it is most economical to drill a well with a rig that does not have significant excess capacity. The cost to the operator for the use of a rig is closely related to its depth capability, so the cost of a well is lowest if a rig is used that matches the requirements of the job. For example, a rig capable of drilling to 20,000 ft should not be used to drill a well to only 7,500 ft, except under special circumstances such as when demand for rigs is high and the larger rig is all that is available. In most cases, though, the well can be scheduled for a time when a more closely matched rig is available.

Of course, rigs are not matched precisely to each small increment of capacity required. Depending on market conditions, contractors will use rigs of varying depth capacities for a given well program. For instance, bids on a group of 7,500-ft wells might be made by contractors offering rigs with depth capabilities up to 12,000 ft.

The same approach is used in the application of offshore rigs, but the water depth capability of the rig is usually a more important criterion than is its drilling depth capacity. The cost of an offshore rig is related to the water depth in which it can work, and it is best to use a rig that is closely matched to the water depth in which it will be working.

For instance, a floating mobile offshore drilling unit outfitted to drill in up to 1,000-ft water depths would not be used to drill a well in 100-ft water depths if it could be avoided. The cost would be high for the floating unit, and a less-expensive jackup rig (bottom supported) could easily handle the water depth involved.

Again, there are exceptions to this general rule—probably more than in the case of land drilling. There are fewer offshore rigs available than land rigs and, depending on availability, offshore units may be used to drill in water depths considerably below their capability. This is the case when the supply of offshore rigs exceeds the demand. Often, offshore drilling contractors would rather have a rig working than idle, even if it is necessary to bid less for the shallower-water contract than the same unit would command in deeper water.

Much of the cost of owning and operating an offshore drilling rig continues whether or not the rig is working. If it is working, even below its maximum capacity, at least some revenue is being generated.

Land and offshore

The many sizes and types of rigs used for drilling oil and gas wells can be broadly classified as either land rigs used to drill wells on land or offshore rigs used to drill in a marine environment. Within each of these main classifications are many sizes and subtypes.

In the case of a land rig, the term *rig* is used to describe all of the items of machinery assembled for drilling, including the mast, drawworks, rotary, pumps, and mud system. Even on a single rig, many of these components can be supplied by different manufacturers. Sometimes the rig is assembled by a firm that does not supply any of the components.

The common use of the term *rig* in the case of offshore rigs has a slightly different meaning. The drilling equipment is basically the same, but it is mounted on a marine vessel that serves as a working platform during drilling. In common usage, rig includes both the drilling machinery and the marine vessel on which it is mounted.

Rig costs. There is almost as wide a range of costs as there is of rig types. At the low end of the scale are truck-mounted rotary rigs for land drilling that can be used to drill shallow oil or gas wells or to perform maintenance and repair on existing wells. The most expensive land drilling rigs are those with extreme depth capabilities, say 30,000 ft or more. In 1981, the cost of a typical superdeep rig ranged up to nearly $10 million. Generally, the deeper a land rig can drill, the more costly it is to build. However, special equipment can have some effect on overall rig cost, resulting in minor exceptions to this rule.

Offshore rigs, because a massive platform must be built to support the drilling machinery, cost several times more than a land rig. In 1981, for example, some offshore rigs ranged in cost from about $20 million to $115 million each. In general, cost of an offshore rig increases as the water depth in which it can operate increases. However, there are two fairly well-defined cost ranges, one for bottom-supported rigs and one for floating rigs.

In the bottom-supported category, a large jackup rig costing about $45 million was ordered in 1981. The practical water depth limit for drilling with a bottom-supported rig is about 300 ft.

For deeper waters, a floating drilling platform is required and the cost escalates rapidly. Several semisubmersible offshore rigs (a type of floater) were ordered during 1981. The cost of each rig ranged from $80 million up to $115 million for a specially designed, self-propelled semisubmersible outfitted for operation in severe environments. A number of semisubmersible offshore rigs were ordered during 1981, costing around $100 million each.

To compare the costs of drilling rigs, it is necessary to know the year in which a rig was ordered or built. Rig building costs have risen dramatically in the past decade as a result of increases in the cost of individual rig components. The price of steel is an important key to how rig costs change, especially in the case of offshore rigs because the vessel on which drilling equipment is mounted consists of thousands of tons of steel.

Rental rates. Within a given type of rig and a range of drilling depth capability, rig rental rates—the amount the contractor charges the well owner—depend on geographic location, the availability of rigs, and special equipment the rig may include. When the demand for rigs is high relative to the supply, more rigs are rented to the operator for drilling on a day-rate basis. When many rigs are idle, this trend shifts and contractors may be more willing to bid on a footage-rate basis in order to be more competitive. In general, footage rates are used more often on wells of modest depth where drilling conditions can be predicted accurately.

In shallower drilling, some turnkey work is done in which the contractor will drill and complete the well for a lump sum amount. Combinations of footage rates and day rates may also be used on a single well.

The geographic location of a well is a key factor in setting drilling rates. A contractor builds rigs suitable for the type of drilling he expects in his operating area and is reluctant to compete, for instance, in the Gulf Coast U.S. land drilling market if his primary operating area is in the Rocky Mountains. Even if the Gulf Coast area is very active, a Rocky Mountain area contractor must compete where he is despite conditions in other locations.

Rig rental rates vary widely with rig rating and geographic area. They may change from month to month in a given area as market conditions change. An example of recent spot rental rates in the Texas Gulf Coast area is $6,000–7,000/day for a land rig capable of drilling to 10,000–15,000 ft, excluding fuel.

Offshore rig rental rates are affected by the same factors as land rig rates. Daily rental rates—the basis on which most offshore drilling is done—are much higher for offshore units. The reasons are obvious: capital investment is several times greater for an offshore rig because (1) the vessel on which the drilling equipment is mounted accounts for most of the cost and (2) offshore operating costs are much higher.

The rental rate includes operating costs, insurance, debt service, and a profit. One rule of thumb is that the daily rental for an offshore mobile rig will be $1,100–1,400/day per million dollars of rig cost. For example, a deep-water semisubmersible that cost $80 million to build and equip might rent for $88,000–112,000/day. Nevertheless this is only a guide and the offshore mobile rig market, or rig availability, is the key factor in setting rig rental rates. The market can be very localized and is related to the water depth capability of the rig.

If the demand for shallow-water rigs in the U.S. Gulf Coast, for instance, is low and rigs are idle, those rigs may not be capable of operating in another offshore area where demand is high. So rates are set in each area according to the demand in that area for a particular type of rig.

When demand for rigs is high, newer rigs also tend to set the rates. An older rig, though it cost much less than a new rig of the same type and capacity, might command the same daily rate as the newer rig. This would mean the rate would be much higher for the older rig than that determined by the rule of thumb mentioned earlier. In periods of slack demand, the new rig would probably work for a lower rate.

Rotary rigs

There is a variety of types of rotary rigs. Many early rigs were built around what is now called a standard derrick in which the mast was assembled on the well location and remained after the well was completed for use in well workover and repair operations. Most rotary rigs used today, however, are moved from one well to another and are used only for drilling. Though even the large rigs are movable, they must be transported in a number of packages or modules. There are also truck-mounted rotary rigs used to drill shallower wells and to perform remedial work on deeper holes.

In late 1981, about 4,500 rotary rigs were operating in the U.S. An additional 1,700 were operating elsewhere.[1]

Slant rigs. Special adaptations of the rotary rig have been used, including the slant rig. Slant rigs are used when it is necessary to drill a well at an extreme angle from the vertical. By tilting the mast or derrick, the hole is deviated from vertical at a shallower depth and with more ease.

Few slant rigs are in use, but they do have application, such as drilling from offshore platforms.[2] Most wells drilled from offshore platforms are drilled directionally to reach out from the platform and penetrate the producing formation in a prescribed pattern at widely spaced points. On land, the rig is moved to a point on the surface above the desired location of the bottom of the well. But in drilling offshore, it is much more practical to drill a number of wells from one location: the platform.

The problem is that the bottom of the wells must be spread out to cover the producing field, just as is the case on land. The answer is usually to begin the hole vertically. After the hole has reached a prescribed depth, it is angled off from that depth so the bottom of the hole will reach the desired lateral displacement from the platform.

This technique normally is done with a conventional rig with a vertical mast. In some cases, however, it is difficult to reach very far from the platform before intersecting the producing formation if this method is used. High angles are necessary in such wells, and additional platforms may have to be installed to develop the field adequately.

Highly deviated or high-angle holes are difficult to drill and may present problems during the producing life of the well. Wear in the casing from drill pipe and tools rubbing on it during drilling, for instance, increases the chance of casing failure. Stress imposed on pipe and other drilling equipment that must traverse the curves in a deviated hole may also cause failure.

When using a slant rig, the operator can start the hole at an angle off vertical. With the same angle buildup, the well can reach a greater horizontal distance before intersecting the producing formation. This approach is especially useful in developing shallow fields from offshore platforms. It may mean fewer platforms are needed, so some of the problems of drilling high-angle holes that are begun vertically are reduced.

Aside from the fact that the mast is slanted from the vertical and is modified somewhat from conventional rigs, most rig equipment is similar to that used on other rigs.

The technique could also be used on land, but to date it has seen limited application. Most directional drilling is still done by starting the well vertically using a rig with a vertical derrick.

Other approaches. Other specialty rotary rigs have also been developed but have not been widely used. The automatic rig, developed during the late 1960s and early 1970s, features automated pipe-handling equipment aimed at speeding the drilling operation and reducing the number of crewmen required. This rig has done a considerable amount of test drilling and is available to the industry.

Another specialty rig that has been tested is the Flexidrill. Instead of rigid steel pipe, this rig uses a flexible drilling hose as a drill string. The flexible hose contains electrical conductors that provide power to drive a motor at the bottom of the drill string to rotate the bit.

Despite extensive efforts to develop more efficient ways to drill, the conventional rotary method—turning a string of steel pipe at the surface with a bit attached to the bottom—still does the bulk of the oil and gas drilling in the world.

Downhole drilling motors. One modification of the conventional rotary drilling method that has been used more widely in recent years is the downhole drilling motor. Russia pioneered downhole-motor drilling, and an estimated 80% of Russia's wells are currently drilled with this tool.

When a downhole motor is used, it is attached to the lower end of the drill string; then the bit is attached to the bottom end of the downhole motor. The drilling motor rotates the bit, rather than the bit being rotated by the drill pipe string.

Conventional rotary rigs are used when drilling with downhole motors. All of the major rig equipment is the same as that used when rotating the bit with the drill pipe. Even the rotary is used to turn the drill string slowly when drilling with a downhole motor. This prevents the pipe from sticking in tight places in the hole, but it does not provide the main rotating power for the bit. Instead, power to turn the bit is provided by the downhole motor. The flow of drilling fluid through the motor—the drilling fluid that is circulated when drilling in the conventional way—powers the drilling motor. The downhole motor converts this hydraulic horsepower into mechanical horsepower to rotate the bit.

The capacity of the rig's mud pumps is critical when drilling with downhole motors. In effect, drilling is being done with these pumps instead of with the rotary.

At one time downhole motors were used primarily to change the direction of the hole when a directional well was being drilled. Now they are being used increasingly for drilling the straight sections of directional wells and for drilling vertical holes. Downhole motors are still used for only a small portion of drilling in the U.S., but one estimate is that drilling motors could be drilling 15% of the well footage in the U.S. by 1985, up from an estimated 1% today.[3]

Source of power

Rotary drilling rigs are also classified by the source of their power. Early rotary and cable-tool rigs were powered by steam. A few steam-powered rigs continued in use until the early 1960s, but by that time most had been replaced by power systems using internal combustion engines fueled by either diesel or natural gas/gasoline.

Today, the primary source of power for rotary drilling rigs is diesel, accounting for about 80% of all rig power in the U.S. The use of natural gas or gasoline to drive the rig prime movers has declined steadily in the past 20 years as diesel has taken over an increasing share.

For many years, power from either steam or diesel prime movers was transmitted to the drilling equipment through belts, chains, or gears and a compound shaft. A rig thus designed is referred to as a *mechanical drive rig*.

Another type of rig power system is diesel-driven DC generators that provide electricity to drive a DC motor on each equipment component. Many of today's rigs use this power configuration. There have been many advances in rig power systems aimed at providing more flexibility for drilling operations, more control, and more efficiency.

Significant in this respect was the development of the variable-speed drive systems using silicon-controlled rectifiers (SCRs) in the mid-1960s. In this power system, diesel-powered alternators supply AC power to a common bus. Then rectifiers convert this power to DC current to drive DC motors that operate the drawworks, rotary, and mud pumps. Auxiliary rig services (lighting, for example) that require AC power are supplied directly. This power system has increased efficiency, fuel savings, and flexibility of operations.

The diesel-electric rig, whether straight DC drive or an SCR system, is the most common rig used today for oil and gas well drilling. More use is expected of the SCR system in the future.

Although steam had virtually disappeared from the scene, it is the source of power for a new rig completed in 1981. Natural gas is used as fuel to generate the steam. It is unlikely, however, that the use of steam power plants will stage a comeback. The more likely trend is that a growing number of rigs will use SCR power systems whose prime movers are powered by diesel and more mechanical rigs will be converted to diesel-electric configurations.

Offshore rig types

Power systems and most rig components are similar whether a rig is drilling on land or offshore. The same criteria are used to select the rig in either case.

After the hole is started in the ocean floor, it is drilled in much the same way as it would be on land. Special techniques and equipment are required to extend the pipe and other equipment from the ocean floor to the above-water drilling platform where the well can be maintained and serviced or to complete the well and install a wellhead on the ocean floor.

While drilling from a floating offshore drilling unit, it is necessary to provide a means for drilling fluid to return from the hole to the drilling vessel. To provide that return flow path and to guide the drill pipe into the hole on the ocean floor, floating offshore drilling is done using a marine riser system. The riser is connected to the well control equipment (blowout preventer) on the ocean floor and to the drilling vessel at the water's surface. Since floating offshore drilling vessels are constantly moving, the riser system includes components to allow some motion of the vessel, both vertically and horizontally, without causing the riser to become disconnected from either the sea floor well control equipment or the drilling vessel.

The most unique aspect of offshore drilling is the support platform needed for the drilling equipment. A variety of marine vessels and platforms have been designed to provide this support in almost all ocean environments.

Fixed-platform rigs. The two major classifications of these structures are mobile and fixed. Fixed platforms (Fig. 2–1) are supported on the ocean floor, normally by driving piles through the platform legs into the sea floor. Fixed platforms are installed after a field has been discovered by an exploration well drilled with a mobile offshore drilling rig.

Development drilling usually takes place from a fixed platform. When all of the wells needed to develop the field are drilled and completed, the drilling equipment is removed and production equipment is

Fig. 2-1 Fixed drilling/production platform. (courtesy Oil & Gas Journal)

installed to handle and process oil and gas produced from the field. Oil and gas leave this platform after processing and are moved to shore, normally through a submarine pipeline.

Drilling rigs used to drill these development wells from a fixed platform are very similar to rigs used to drill on land. Depending on the size of the platform, however, some of the equipment may be mounted on a tender or barge tied alongside the platform. Electrical lines and piping then connect this tender-mounted equipment to the rig floor.

Platform rigs, as the rigs mounted on fixed platforms are called, are used to drill a number of wells from a single platform. Several platforms with a group of wells drilled from each may be required to develop a single offshore field adequately. Platforms of this type are installed in water depths ranging from a few feet to more than 1,000 ft. In addition to fixing the platform to the ocean floor with piles, other fixed platform designs include a tethered platform for deep water, which is anchored to the ocean floor with cables, and a tower design, which is anchored by a base and a huge universal joint.

Two giant platforms show the extent of the industry's capability. One, installed in 1,025 ft of water in the Gulf of Mexico, was fabricated in three sections: a base, a mid-section, and a top section.[4] The platform's top section, 530 ft high and weighing 11,000 tons, was set about one year after the base section was installed. A 2,000-ton deck and two platform drilling rigs were then installed on the top section. Shell, operator of the field, said the cost of the project, including pipelines, was about $800 million. It estimated the platform will produce 100 million bbl of oil and 500 billion cu ft of gas over its lifetime.

Another deep-water producing platform set in 935 ft of water by Union Oil is said to be the largest single-piece offshore drilling/production platform jacket.[5] The 25,000-ton jacket is 952 ft high and is

Fig. 2-2 Barge rig for inland waters. (courtesy Oil & Gas Journal)

scheduled to begin production in 1985. The structure was towed to the site on a specially built 650-ft barge, launched, flooded, righted, and then installed on the ocean floor. Production is expected to reach 25,000 b/d of oil and 96 MMcfd of gas.

Mobile rigs. Exploratory wells are drilled offshore using mobile rigs, rotary drilling rigs mounted on a marine vessel. Mobile drilling rigs are either bottom-supported or floating units.

Barge rigs (Fig. 2–2) and submersible rigs are used in shallow waters, including inland waters. They are towed to the drilling site and ballasted to rest on bottom to provide a firm base for drilling. Of the worldwide offshore rig fleet, about 56 rigs are of these two types.[6]

For medium water depths, the jackup mobile rig (Fig. 2–3) is widely used. It is a bottom-supported platform but is moved from one exploratory well to another, distinguishing it from a fixed offshore platform. The vessel consists of a hull on which the drilling and auxiliary equipment is mounted and three or more legs that can be jacked up and down. The vessel is towed to location with the legs jacked up above the hull.

Fig. 2–3 Jackup offshore mobile drilling rig. (courtesy Oil & Gas Journal)

When the vessel is at the drilling site, the legs are jacked down until they rest on the ocean floor. Further jacking down of the legs lifts the hull out of the water and provides clearance between the bottom of the hull and the water's surface. A detail of a jackup rig jacking system is shown in Fig. 2-4.

Jackup rigs come in all sizes and configurations, and about 358 are in service around the world. They are a popular rig, as evidenced by the number on order at the end of 1981, because they can be used in a wide range of water depths. The water depth limit for these units is generally considered to be about 300 ft, although a jackup has been designed for use in 400 ft of water.

A modification of the conventional jackup is the mat-supported jackup. In this version, the bottom end of each leg is connected to a base, or mat, that rests on the ocean floor, providing additional contact area to support the platform.

For water depths beyond the capability of bottom-supported units, floating drilling rigs are used. The most common of these is the semisubmersible (Fig. 2-5). Normally a vessel with a roughly rectangular-

Fig. 2-4 Jacking system used on offshore jackup rigs. (courtesy Oil & Gas Journal*)*

TYPES OF RIGS 41

Fig. 2–5 Semisubmersible offshore rig under tow. (courtesy Oil & Gas Journal)

shaped hull, it is anchored over the drilling site with a number of anchors and connecting chains spread out around the vessel in a well-designed pattern. One of several anchoring windlasses on a semisubmersible offshore rig is visible in Fig. 2–6.

These vessels have been used to drill in record water depths of several thousand feet and can operate in severe ocean wave and wind conditions. They are available in a variety of configurations.

Most of these vessels are towed to the drilling site by tugs, but a few have been outfitted with propulsion systems and can travel under their own power.

A further development of the propulsion system on a few of these vessels provides automatic station-keeping, eliminating the need for anchors. By coupling the propulsion system with a system to monitor the rig's location precisely, the vessel is kept over the well automatically. Any tendency to drift away from the hole is detected by the monitoring system, which signals an appropriate thruster to operate to keep the vessel in the designated spot.

Fig. 2–6 One of semisubmersible offshore rig's marine anchoring windlasses, foreground. (courtesy Oil & Gas Journal)

TYPES OF RIGS 43

Fig. 2–7 Drillship, another type of mobile offshore drilling unit. (courtesy Oil & Gas Journal)

Few wells are drilled using this automatic station-keeping system because operating expense is high. But the industry has the capability.

Another type of floating deep-water drilling unit is the drillship (Fig. 2–7). The drilling equipment is mounted on a ship-shaped hull, often equipped with a propulsion system to eliminate the need for tugs when moving from one location to another. These vessels are used much like the semisubmersible floating drilling unit; they are anchored on location and the drilling operation is conducted the same as on a semisubmersible rig.

Drillships have been popular for remote exploratory drilling because they normally have a large storage capacity for pipe and supplies and can move long distances between wells in a shorter time than other types of offshore drilling units. They make up about 10% of the world's offshore drilling fleet.

Drilling from a floating platform poses several unique problems not found in drilling either on land or with an offshore rig mounted on a fixed platform.[7] One of the most important is the fact that the floating rig is always in motion. The severity of this motion varies with the weather or sea state. Both horizontal and vertical motion of the vessel and its drilling equipment must be minimized or compensated for in floating offshore drilling.

To make hole, the bit must be in contact with the bottom of the hole while it is rotating with weight applied. There is no penetration if the bit is above the bottom of the hole, spinning. But unless compensated for, this would be the position of the bit much of the time when drilling from a floating vessel. A small amount of vertical motion will lift the bit from the bottom of the hole, even allowing for the stretch of the drill string.

To compensate for this motion and to keep the bit on bottom with the desired amount of weight, a common approach is to use a slip joint or bumper sub in the drill string that serves as a sort of shock absorber. It offsets the motion of the rig relative to the drill string and keeps the bit on the bottom of the hole.

Another approach used to counter vertical motion of the vessel is a tensioning system that controls the tension in the drilling line which suspends the drill string from the derrick. The aim of this system, as well as the bumper sub system, is to keep the bit on bottom drilling even though the vessel and the drilling equipment is moving vertically (heaving).

Horizontal motion of a floating drilling vessel must also be allowed for. Since a riser connects the floating drilling vessel to the sea floor, it is necessary to keep the drill string as near vertical as possible. If the vessel moves horizontally, the drill string and riser can bend, causing excessive wear and even failure of the riser, drill pipe, or other equipment. Limits are usually imposed on how much horizontal displacement is permissible. If sea conditions cause this to be exceeded, it may be necessary to disconnect from the well and suspend operations until conditions will allow maintaining the permitted horizontal displacement.

Reliable information on the position of the vessel relative to the sea floor location of the well is necessary to stay within horizontal displacement limits.

Another unique aspect of floating drilling is that the well control equipment is located on the ocean floor. Hydraulic and electric systems are necessary to operate this control equipment from the rig floor, and the complexity of the operation is increased. Because this equipment is located far from the rig floor in a marine environment, starting a well from a floating drilling vessel is different from other types of drilling. More detail on this operation will be given in chapter 6.

Moving rigs

Virtually all drilling rigs used today are mobile, whether used offshore or on land. The degree of mobility varies; some may move virtually intact, while others require several loads by truck or barge to move from one drilling site to another.

Land rigs are normally moved in packages by truck and assembled at the well site. The number of modules or truck loads comprising a land rig varies widely. Typically, though, the drawworks is one load, the mud pumps are one load, mud tanks and other parts of the mud system may make up several loads, and the mast is moved as a unit. The mast is normally transported in a horizontal position. Various other parts of the rig require additional loads, the number depending on the size of the rig and the equipment it contains.

Over the years, considerable effort has gone into designing rigs to be as compact as possible, requiring the fewest loads to move. Much work has also gone into designing the individual modules so assembly and hookup at the rig site is easy and quick.

Some small rigs are truck-mounted, and the bulk of the equipment remains on the truck when the rig is erected for drilling. These rigs can usually be rigged up—made ready to drill—very quickly, often within a few hours.

In special cases, such as in desert areas where a relatively short move may be required and no roads exist, a rig may be moved with very little disassembly and with the mast in the vertical position. These rigs are mounted on large, rubber-tired bogie units that can be pulled to the new location by heavy trucks or tractors.

Other special situations call for unique transportation solutions. In the Arctic, for example, permafrost must be traversed with care. The weight of a drilling rig moved in a conventional way could damage the terrain severely. One approach to this problem is an air-cushioned vehicle for moving equipment. The air-cushioned vehicle supports its load by maintaining air pressure between the bottom of the unit and the ground. A skirt confines the air under the bottom of the vehicle. This approach has not been used extensively, however. When drilling must be done in these areas, gravel roads are carefully built over which to move equipment. Properly constructed, these roads prevent damage to the sensitive environment.

Remote jungle areas spawned the design of another type of drilling rig: the helicopter-transportable rig (Fig. 2–8). A number of rotary drilling rigs have been specifically designed and built with the load-carrying capacity of large helicopters as a design criterion. Obviously,

46 FUNDAMENTALS OF DRILLING

Fig. 2–8 Helicopter-transportable rig being assembled. (courtesy Oil & Gas Journal*)*

these rigs must be disassembled into smaller packages than if they were to be moved to location by other means.

Most offshore rigs are moved from one location to another by tugs. Rig tows of several thousand miles are not uncommon. Many are built in the U.S. Gulf Coast construction yards, for example, and are put to work in virtually all areas of the world. The expansion of the mobile drilling fleet over the last 30-plus years has brought a corresponding growth in the need for rig-towing services.

A few offshore rigs, as mentioned earlier, are self-propelled and move without the aid of tugs.

A novel approach to offshore rig moving was used in 1980 when two 180-ft tall sails were attached to the legs of a jackup rig to assist in a tow of over 2,500 miles.[8] The same firm had used the method before, but this technique is rare. The sails can only supplement power provided by the tow boats and cannot provide all of the power needed.

Workover rigs

Although not normally designed for conventional drilling operations, well-servicing and workover rigs (Fig. 2-9) perform vital functions in the oil and gas industry.

The type, cost, and size of these rigs varies widely, and it is impossible to describe an average rig in the broad category of workover and well servicing. Nevertheless, these rigs are often loosely classified according to the type of work they do: well workover, well service or maintenance, and well completion.[9]

Workover rigs often are able to perform all of the functions of a conventional drilling rigs: hoisting, rotating a drill string, and circulating drilling fluid. As a result, some of the larger units may be used to drill shallow depth wells. Performing a workover of an existing well, however, is their main function. A workover operation usually consists of making changes in the downhole equipment in a well or modifying the producing interval. It may mean recompleting a well to produce from a different zone than the original producing zone or repositioning well equipment within the original producing zone.

Well servicing rigs normally perform operations involving well maintenance, such as replacing the rods used to pump an oil well or swabbing fluid from the well to reestablish production. The jobs normally require a shorter time than a well workover since no significant change is being made in wellbore equipment or the producing zone. Well servicing rigs, as well as workover rigs, are highly mobile. The time

48 FUNDAMENTALS OF DRILLING

Fig. 2-9 Service/workover rig with mast in traveling position. (courtesy CMI Corp.)

required for both types of operations is typically much shorter than the time required to drill a well.

Another function of this general type of equipment is well completion. Especially in the case of deep holes where a large, costly drilling rig is required for drilling, the rig may be moved off the well when the last string of casing is installed and cemented and a completion rig brought in. Releasing the large rig, which has a high daily rental rate, when its capacity is no longer needed can reduce the well cost. The lower-cost completion rig is brought in to install producing equipment in the hole and to get the well ready for production.

Any one rig of the general types described may perform a workover, well maintenance, or completion. But many are best suited for one of these operations. In the U.S. alone in mid-1981, almost 6,000 workover/service/completion units were available.[9] It was estimated that in 1981 about 1,245,000 well service, workover, and completion jobs would be performed. Nearly 1,045,000 of these were well service operations, 149,000 were well workovers, and 51,000 were well completion jobs.

The size of these rigs covers a broad range. Units with a large lifting capacity are needed when pulling heavy tubing from deep wells, for example. And a unit with a mast tall enough to permit pulling three joints of tubing at a time rather than one would be desirable in a deep well.

References
1. "Hughes Rig Count." *Oil & Gas Journal*. (4 January 1982), p. 174.
2. Dodds, Robert G. "Slant Rigs Offer Wider Reach from Offshore Platforms." *Oil & Gas Journal*. (8 May 1978), p. 211.
3. Moore, W.D., III. "Drilling Motors Evolve to Valuable Drilling Tool." *Oil & Gas Journal*. (9 March 1981), p. 75.
4. "Cognac Platform Work Advancing." *Oil & Gas Journal*. (14 August 1978), p. 50.
5. "Union Installs Big Jacket off Texas." *Oil & Gas Journal*. (13 July 1981), p. 41.
6. Moore, W.D., III. "Offshore Drilling Maintains Fast Pace." *Oil & Gas Journal*. (3 May 1982), p. 143.
7. Harris, L.M. *An Introduction to Deepwater Floating Drilling Operations*. Tulsa: PennWell Publishing Co., 1972.
8. "Jackup Working off Canada after Sail-assisted Tow." *Oil & Gas Journal*. (13 October 1980), p. 79.
9. Hyde, R.B. "Projections and Trends in Well Servicing." *Well Servicing*. (September–October 1981), p. 42.

3 Major Rig Components

THE rotary rig must perform three basic jobs during drilling. All types of rotary rigs, whether designed for deep or shallow drilling, whether mounted on a marine vessel for offshore drilling or drilling on land, have similar equipment components that perform three basic tasks:

1. a drawworks that, through a system of cable and pulleys mounted in the derrick, lowers drill pipe and casing into the hole and hoists the drill pipe string out of the hole
2. a rotary table and related equipment that rotates the drill string to turn the bit on the bottom of the hole
3. a mud system that circulates drilling fluid to the bottom of the hole and back to the surface to remove cuttings from the bottom of the hole, to cool and lubricate the bit, and to control downhole pressures.

There are many individual items of equipment within each of these groups, but these three systems, along with the primary power source, are the heart of any rotary drilling rig (Fig. 3–1).

Additional equipment is needed for auxiliary services such as lighting. A control system and instrumentation to monitor drilling conditions such as weight on bit, mud pump pressure, and rotary speed are also key components of the rotary rig. These systems vary in complexity, depending on rig size and type. Also, on critical wells or where conditions cannot be predicted accurately, much more sophisticated monitoring equipment may be brought to the well site.

In remote areas, additional services may be required, such as sleeping quarters for crews. Storage is also required for water, fuel, and various drilling fluid components. The amount of storage depends on the location of the rig and on expected drilling conditions.

MAJOR RIG COMPONENTS   51

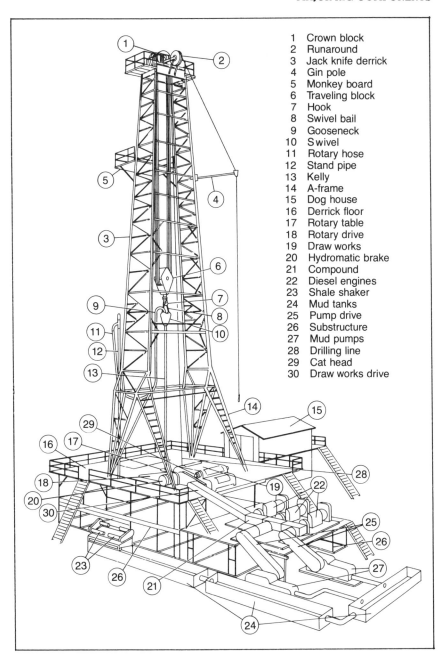

1 Crown block
2 Runaround
3 Jack knife derrick
4 Gin pole
5 Monkey board
6 Traveling block
7 Hook
8 Swivel bail
9 Gooseneck
10 Swivel
11 Rotary hose
12 Stand pipe
13 Kelly
14 A-frame
15 Dog house
16 Derrick floor
17 Rotary table
18 Rotary drive
19 Draw works
20 Hydromatic brake
21 Compound
22 Diesel engines
23 Shale shaker
24 Mud tanks
25 Pump drive
26 Substructure
27 Mud pumps
28 Drilling line
29 Cat head
30 Draw works drive

Fig. 3-1 Drilling rig components. (courtesy PennWell Publishing Co.)

On offshore rigs, a much wider range of equipment is needed. Anchoring systems require chain storage; crew sleeping quarters and mess facilities are needed. Special equipment for the drilling operation must also be aboard, such as diving systems.

But as far as drilling the hole is concerned, all rigs—including offshore drilling rigs—have the three basic equipment groups in common: hoisting, rotating, and pumping.

Drilling and completing a well also requires equipment that is not part of the conventional rotary rig. Cementing casing in place calls for cement pumping trucks, mixers, and additional piping. Logging requires special trucks outfitted with a winch to lower the logging tool into the hole and sophisticated instrumentation to record data gathered by the logging tool. These services are performed by firms other than the drilling contractor that bring the necessary equipment and specially trained personnel for a specific job to the wellsite. When that job is complete, the specialized equipment and crews are removed and drilling continues.

A wide variety of these services is required, depending on the type of well, the amount of casing required, how much logging and testing is done, and whether any problems develop during drilling. For example, if equipment is lost or dropped in the hole, it must be removed before drilling can continue. A small piece of the bit, for instance, may break off while drilling, or several thousand feet of drill pipe may drop to the bottom of the hole if the drill string parts. The job of fishing to remove this equipment from the hole calls for a firm with special tools and specially trained crews.

Offshore rigs, depending on their size and capability, often have cementing and logging equipment permanently mounted on the vessel. When these services are required, a special crew will perform the job, using this permanently mounted equipment.

Hoisting

Two key items involved in lifting or lowering drill pipe or casing are the derrick, or mast, and the drawworks. Early derricks were built in place. After they were used to drill the well, they were left standing to be used for workover and maintenance of the well during its producing life. Some of these remain in old oil fields around the country, but most have been removed.

Derrick. Most modern land rig derricks are of the jackknife type. They are transported to location in a horizontal position and are then raised on a base to the vertical at the wellsite. The derrick must be designed to support the weight of the drill pipe string that will be needed while drilling the deepest well for which the rig is designed. It should also be capable of an *overpull*, a force in addition to the weight of the drill string that may have to be exerted to pull the drill pipe through a tight spot in the hole.

When pulling the pipe or lowering it in the hole, the total weight of the pipe string and any overpull force is transferred to the legs of the derrick. A system of pulleys at the top of the derrick—the crown block—transfers this weight from the drawworks through the drilling cable to the derrick.

Often in drilling deep wells the weight of casing installed in the well will exceed the weight of the drill string used at any time. This must be determined in planning the well. A rig must be selected with a derrick that can support the expected casing weights.

In addition to sufficient strength to support the weight of drill pipe and casing, derricks on modern rigs are designed with sufficient height so pipe can be pulled in triples rather than singles. A single is one joint or length of pipe with a connection on each end. As the well is drilled and the bit makes its way lower in the hole, a joint of drill pipe is added at the surface to lengthen the drill pipe string. When the drill string must be pulled from the hole, it is much faster if only one out of every three of these screw connections must be disconnected and reconnected. So the pipe is pulled in stands—sections consisting of three 31-ft joints.

These stands are racked vertically in the derrick as they are disconnected. Then when the drill string is lowered back in the hole, the procedure is reversed. Each stand is reconnected and the drill string is lowered into the hole.

The requirement for racking these stands of drill pipe in the derrick during a trip (pulling the bit from the hole and lowering it back to bottom) is another criterion for derrick design. The derrick must be designed, for instance, to withstand the wind load that would be imposed on the derrick and on the drill pipe when the drill pipe is standing vertically in the derrick.

The rig base, or substructure (Fig. 3–2), on which the derrick and other equipment is mounted must also be designed to support the weight of the drill pipe when it is stacked in the derrick. In the case of a deep hole—20,000 ft, for example—when all of the pipe is out of the hole, over 200 stands of drill pipe will be racked in the derrick. Assuming 4½-in., 16.60-lb/ft drill pipe is being used, the total weight of the pipe in the derrick is over 300,000 lb.

In some wells the weight of one of the casing strings may be greater

54 FUNDAMENTALS OF DRILLING

Fig. 3–2 High-floor substructure being assembled in rig-up yard. (courtesy National Supply Co., an Armco Group)

than the weight of the drill-pipe string. This must be considered when selecting a rig. For instance, if the well plan calls for 7,000 ft of 10¾-in., 55.5-lb/ft casing to be installed, the total load on the derrick while lowering this casing into the hole would be almost 390,000 lb, neglecting the buoyancy effect.

Drawworks. The drawworks is a revolving drum around which the drilling line—steel cable—is wound (Fig. 3-3). The line unwinds from the drum up through the pulleys or sheaves in the crown block and down to the sheaves in the traveling block when lowering the pipe into the hole. When pulling drill pipe from the hole, the system reverses direction. It is equipped with a brake that can stop and hold the drill string at any point in the hole.

The drawworks and its power source are sized according to the depths to which the rig will be drilling. The drawworks must be able to pull any anticipated weight of drill pipe from the hole and often must be able to

MAJOR RIG COMPONENTS 55

Fig. 3–3 Electrically driven drawworks (background) and rotary table (foreground). (courtesy National Supply Co., an Armco Group)

exert an overpull if the pipe becomes stuck while being pulled from the hole. Other duties of the drawworks include raising and lowering the derrick during rig-up and rig-down and making up and breaking out threaded pipe and casing connections.

In a mechanical-drive rig, power for the drawworks is transmitted from the rig's engines through belts, chains, or a compound shaft (Fig. 3–4). With an electric rig, the drawworks is powered by a DC electric motor. The electric motor may be driven directly by a DC generator powered by a diesel engine or it may be driven by DC current from a silicon-controlled rectifier (SCR) unit, which in turn is supplied with AC power by AC generators driven by the rig's engines.

In the case of an electric rig, the drawworks may be driven by one, two, or possibly three 800–1,000-hp motors, depending on the drilling depth for which the rig is designed. On a mechanical rig, the power available to the drawworks will be similar and will depend on the depth capacity of the rig.

56 FUNDAMENTALS OF DRILLING

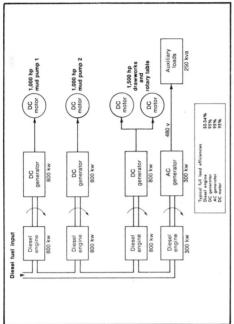
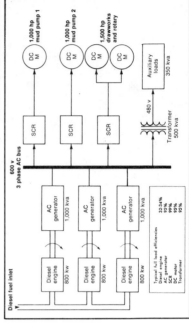
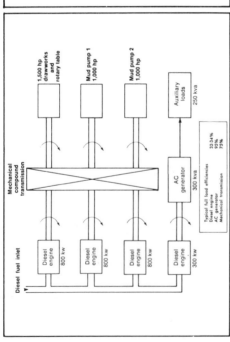

Fig. 3–4 Types of drilling rig power include mechanical (above), DC-DC (above, right), and SCR (right). (courtesy Oil & Gas Journal)

MAJOR RIG COMPONENTS 57

Hoisting equipment, including both derrick and drawworks, is the key criterion in determining the depth capability of the rig. The size and horsepower of other rig components also varies with the depth capacity of the rig, but not in as direct a proportion as the drawworks and derrick must.

Elevators are part of the hoisting equipment and must support the weight of the drill pipe string or casing when lowering into the hole or coming out of the hole. A traveling block (Fig. 3–5) is also part of the hoisting system.

Rotating

In conventional rotary drilling, the rotating force to turn the drill pipe and rotate the bit at the bottom of the hole is provided by the rotary table (Fig. 3–3). An exception is when drilling is being done with a downhole motor. In that case, the bit is rotated by the downhole motor rather than by the rotary table. Even when drilling with a downhole motor, how-

Fig. 3–5 Traveling block. (courtesy National Supply Co., an Armco Group)

ever, the drill pipe is usually turned slowly with the rotary table to keep the drill string free and to prevent sticking the pipe.

The rotary table rotates on bearings and turns the drill string through the kelly, a length of pipe or hollow forging with shoulders on the outside that make it either square or hexagonal. The kelly fits into a matching shouldered hole in the rotary table and is screwed into the top of the drill pipe string. When a new joint of drill pipe must be added because the hole has been deepened, the kelly is unscrewed from the drill string and a new joint of drill pipe is screwed in. Then the kelly is screwed into the top of the new joint of drill pipe and drilling continues.

While these connections are being made on the rig floor, the drill pipe is suspended in slips in the rotary table. Slips support the weight of the drill string while a new joint is being added and the kelly is being screwed into the new joint.

When the connections are made, the drill string is raised slightly to take the weight off the slips, and they are removed from the rotary table.

Fig. 3–6 Two main mud pumps, foreground; mud pits in left background. (courtesy National Supply Co., an Armco Group)

Then the drill string is lowered until the bit is again on bottom and the kelly is in the rotary table, and drilling resumes.

When making a trip out of the hole, the rotary table is used to unscrew the drill pipe connection after the drawworks has been used to break out or loosen the threaded connection. When adding pipe, the rotary screws in the new section; then the drawworks is used to make up the connection, or give it a final tightening.

In an electric rig, power for the rotary is provided by an electric motor. On a mechanical rig, the rotary table is driven by a sprocket and chain arrangement. Typically, power required to drive the rotary is 400–800 hp.

An unusual rotary drive system was incorporated recently in a medium-depth rig.[1] The rotary is drvien by a hydraulic motor connected to a pump with hoses. The pump is driven by the rig's main engine compound.

According to the rig's owner, the system has certain advantages over the chain/sprocket drive system, including the elimination of alignment and wear problems. It is also said to provide a better feel for the drilling conditions being encountered by the bit and drill string downhole. The system includes a bypass arrangement that stops the rotary table when a certain torque is reached in the drill string. This feature can help prevent breaking the drill string and other damage. The system is unique for a larger rig; the chain/sprocket drive is normally used.

Fluid handling

Of the three basic systems, there is more variation in the drilling fluid system than in the hositing or rotating groups. What is used on a specific well is determined by well depth, type of formations to be drilled, expected pressures and temperatures, and other factors.

Each rig has one or more main mud pumps (Fig. 3–6) that circulate the drilling fluid down the inside of the drill pipe, out through the bit, and back up to the surface around the outside of the drill pipe (the annulus). Each rig also has holding tanks into which the mud returns and equipment for removing rock cuttings from the returned mud. In addition to these basic items, a wide variety of mud handling, mixing, and treating equipment is available.

The swivel provides a connection through which mud can be pumped into the drill pipe while the drill pipe is rotating (Fig. 3–7).

The complexity of today's drilling fluid systems, developed to meet a variety of drilling conditions, makes the use of special mud equipment necessary in many cases. Mud mixing equipment ranges from basic to

60 FUNDAMENTALS OF DRILLING

exotic and is needed when adding liquid or solid materials to mud to handle special situations. Weighting material—barite, primarily—is often added to drilling fluid to increase its weight and to enable it to keep gas or liquids present in zones penetrated by the bit from flowing into the wellbore. Once these fluids enter the wellbore, a temporary loss of well control or even a blowout may occur. A weighted drilling fluid is therefore used on most deep and medium-depth wells to control formation pressures.

Other materials are also added to the drilling fluid to meet special needs and to adjust the properties of the fluid, such as viscosity or the ability of the mud to form a cake on the walls of the drilled hole.

Removal of cuttings that the mud has carried to the surface is a key job of the drilling fluid system. If these cuttings are not removed at the

Fig. 3-7 Swivel for rotary drilling rig. (courtesy National Supply Co., an Armco Group)

surface, they are carried back to the bottom of the hole when the fluid is circulated and are reground by the bit, slowing penetration. Also, if cuttings build up in the mud, the drilling fluid's carefully designed properties will be adversely affected.

To remove larger particles from the returning mud stream, a shale shaker is used on most rigs. There is a variety of shale shaker designs, but in each the particle-laden mud returning from the bottom of the hole flows over a vibrating screen that lets the fluid and very small particles pass through. The larger particles are retained on the screen and discarded. Screens of different sizes are used on shale shakers to remove particles of different sizes.

To remove finer particles, desanders and desilters may be used. The amount and type of this mud cleaning equipment used on a given well depends on the type of formations being drilled, well depth, and other factors.

Tanks are provided for settling and for pump suction. The size and number of tanks also depends on factors such as well depth and hole size. There is a recommended relationship between the volume of the hole to be drilled and the volume that must be available in the mud tanks. Other design and operating guidelines help determine how much spare drilling fluid and additives should be available at the rig site in case of unforeseen events such as a well kick.

In addition to steel tanks for settling and a suction reservoir for the main mud pumps, there is usually an earthen reserve pit built at most drilling locations. Cuttings removed from the mud stream by the shale shaker are sent to this pit, which may be used if an emergency situation requires directing the flow of mud or well fluids away from the rig.

Air/foam equipment. Different equipment and techniques are needed when drilling with air or foam. The drilling fluid handling system must be modified. Since air is compressible, air compressors must be used to circulate the drilling fluid rather than pumps. Special considerations are also necessary in selecting bits for air drilling operations.

In the mud-handling system, most of the liquid and solids-handling equipment is not needed. Some water storage capacity and a transfer pump are required. Special manifolds may also be needed to direct the flow of the air drilling fluid.

A rotating drilling head is also normally used when drilling with air. It provides a seal around the rotating drill string, blocking off the annulus between the drill string and casing so air returning from the hole with cuttings can be directed away from the rig.

Compressors used for air drilling are not normally part of a conventional rig but are brought to the drilling site for the operation. They are packaged on a specially built truck and are available in various combinations, depending on the pressures and volumes needed for a particular well. A typical air drilling compressor unit is capable of delivering 400–1,200 cu ft/minute at a maximum pressure of 300 lb/sq in.[2]

Other equipment is also needed for air/foam drilling operations, including:

1. a pump to inject water or soap into the air stream if mist or foam drilling is done
2. various sampling and safety devices installed in the line through which the air or mist returns from the well
3. special instrumentation for monitoring air pressure and other drilling conditions
4. a specially designed bottom-hole assembly to control hole deviation and other hole problems

Safety is a key consideration in air drilling. When gas is encountered while drilling with air, the danger of a downhole fire or explosion exists. Preventing ignition under these conditions is an important consideration during planning and drilling.

The drill string

The drill string consists primarily of joints of drill pipe. Other tools in the drill string include the bit and the drill collars.

The drill string is turned by the rig's rotary table and in turn rotates the bit on the bottom of the hole. On the rig floor, the drill string is connected to the mud system by the rotary hose and swivel. Because it is hollow, the drill string serves as a conduit through which drilling fluid is pumped to the bottom of the hole to remove rock cuttings and to cool and lubricate the bit.

Drill pipe. Drill pipe comes in a range of diameters; in each diameter there are several weights and grades of steel. Most drill pipe ranges in diameter from 2⅛ in. to 5½ in.; 4½-in. diameter drill pipe is a common size.

When rigs are rated as to drilling depth capability, those ratings are often based on the use of 4½-in. drill pipe. When considering a stated rig depth capacity, it is important to know what size drill pipe was used as a basis for determining the rating. For example, a given size derrick and drawworks could suspend a greater length of 3½-in. drill pipe than 4½-in. drill pipe because the larger-diameter pipe weighs more per foot. Within the 4½-in. pipe size, the weight per foot of commonly used drill pipe ranges from 12.75 lb/ft to 20.00 lb/ft. Steels of different strengths are used, depending on drilling conditions expected.

Not all drill pipe is made of steel. Aluminum drill pipe has been used in some drilling, but its use is not widespread. Its advantage is that, for a given length (well depth), an aluminum drill string weighs less than a steel drill string. It has been used, therefore, to permit drilling wells with rigs whose capacity would be exceeded if steel drill pipe were used.

As the well is drilled deeper, joints of drill pipe are added to the drill string on the rig floor. These joints are about 31 ft long with a threaded connection on each end. One end has a male threaded connection, called a *pin*; the opposite end has a female threaded connection, called a *box*. These 31-ft sections consist of two parts: the pipe, or hollow tube, and the threaded connections, called tool joints. The tube is manufactured in one process; then the tool joints are welded on each end.

Drill string forces. In a deep well particularly, tremendous forces are imposed on the drill pipe. The longer the drill string, the greater the tension load on the string. This tension load is greatest at the surface, since all of the pipe below is suspended from the top joint of drill pipe when the bit is not resting on the bottom of the hole.

During drilling, weight is put on the bit to make it bore into the rock; so some of the weight of the drill string is not supported by the top section of the pipe during actual drilling. But this top section must be capable of supporting the drill string when lifting the pipe out of or lowering it into the hole.

Torque, or twisting force, on the drill string can also be great, especially in a hole where the drill string does not rotate freely because of a tendency to stick or because the hole is not straight. When this torque exceeds the strength of the drill pipe, the drill string may break or twist off. Then fishing operations must be conducted to retrieve the portion of drill string below the break. Erosion at a tool joint caused by a leak may weaken the drill pipe and cause it to twist off more readily. Or it may have been stressed in other ways, weakening either the pipe itself or the threaded connection.

Though it is usually not the limiting factor, drill pipe must also be able to contain the pressures imposed by the mud pump when pumping drilling fluid down the drill string. Normally, the strengths obtained when choosing drill pipe with sufficient strength in tension and torsion will ensure that the pipe is strong enough to contain the mud pressure and resist burst. Pressures needed for well treatment after drilling is complete may be greater than pumping pressure during drilling, so treatment pressures must be considered when evaluating the burst strength of a drill string to be used on a specific well.

Normally, the threaded connection is the weakest point as far as containing the drilling fluid pressure is concerned. An improperly made-up connection or a damaged thread area may leak when circulating pressure is reached. Continued leaking can erode the metal in the threaded connection area until the connection fails due to tension or torque.

Drill pipe wear can be a significant factor in failure of the drill string. Some wear is normal, of course, but excessive wear can occur in holes that are not relatively straight and under other drilling conditions. Individual or multiple joints of drill pipe may be worn much more than the rest of the drill string. Frequent inspection is necessary to prevent failure when the pipe is in the hole. Drill string failures can be costly because special tools and procedures must be used to fish broken or parted portions of the string from the hole. Failures are also time-consuming because drilling must be suspended until the hole is clear.

In extreme cases, it may be impossible to remove parts of the broken drill string or other junk from the hole, and the hole may have to be deviated or sidetracked around the junk. The hole may even have to be abandoned if sidetracking is impractical.

How long a drill string will last depends on the type of drilling it is used for and how well it is cared for. For planning purposes, one estimate is that a drill string should drill between 300,000 and 400,000 ft if properly inspected and maintained.[3] This is equivalent to 6–9 years of average drilling.

In mid-1981, a string of 4½-in., Grade E. 16.60 lb/ft drill pipe—a common size and grade—cost about $31/ft, including hard-banding, internal coating, inspection and freight.[3] A 10,000-ft string of this pipe would cost over $300,000, not including any special drill string tools.

Other drill string tools. Besides the drill pipe, there are other components of the drill string. Some are used on all wells; some are used in special situations.

The bit, of course, is used on all wells. Immediately above the bit, a number of drill collars are run. Drill collars provide additional weight on the bit and add stability to the drill string to help maintain a straight hole. The drill collar is an extra-heavy length of pipe with threaded connections on each end that can be connected to the threaded joints on the drill pipe and bit.

Drill collars come in a range of sizes, normally from 4½ in. to 8 in. in diameter. A range of weights is available in each diameter. For instance, a 6-in. outside diameter (OD) drill collar is available in weights from 83 to 88 lb/ft, depending on the inside diameter (ID) of the collar. A number of these collars, about the same length as joints of drill pipe, can be connected at the bottom of the drill string. In the case of the 6-in. 88-lb/ft collar, each weighs about 2,700 lb.

Depending on hole conditions and the drilling plan, other tools may be run within the drill string, including stabilizers and reamers. These have a hole through the center so mud flow can continue from the inside of the drill pipe, down through the tool, and through the bit. Rubbers may be attached to the outside of the drill pipe to prevent wear of the casing when the drill string is rotating inside a casing string.

Other equipment

The preceding is by no means a complete list of all tools and equipment on a conventional rotary rig. Only the main groups or systems have been discussed.

There is also equipment needed for drilling the well that is not an integral part of the drilling rig but is brought to the location when needed to perform a specific operation. When it is time to install a string of casing, a firm specializing in this operation normally brings equipment and crews to the rig to run casing. Special casing-cementing equipment—pump trucks, mixers, bulk tanks—is brought to the rig by a cementing company and is operated by cementing personnel in conjunction with the rig equipment operated by the regular drilling crew.

If equipment is lost in the hole and a fishing job is required, the operation may be performed by the rig crew or a specialist may be called in. A few common items of fishing equipment are normally kept at the rig, but others may have to be brought in for special situations or even made at the rig site.

Logging of the hole, normally after the planned depth has been reached, is also done with special equipment and personnel brought to location for that operation. Other equipment and services such as pipe inspection may be required at any time during drilling.

System efficiency

The key factor that determines how fast a hole can be drilled is the amount of energy (horsepower) that can be delivered by the drilling system to the bit at the bottom of the hole. In a conventional rotary

66 FUNDAMENTALS OF DRILLING

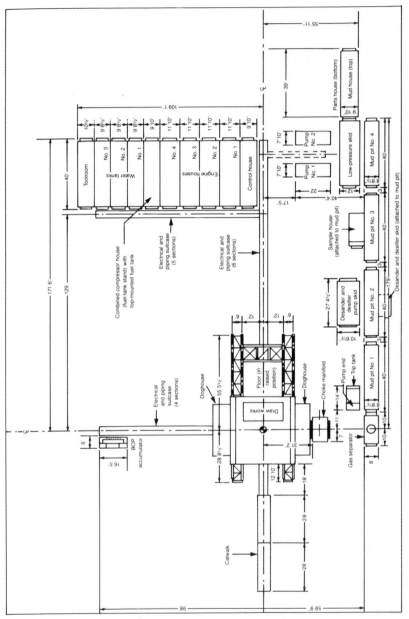

Fig. 3-8 Layout of large land rig. (courtesy Oil & Gas Journal)

drilling system, much of the energy generated at the surface is dissipated before it reaches the bottom of the hole. The rotary drilling system can transmit only about 20–40 hp to the rock, for instance, while a jet-piercing drill might be able to transmit 500–1,000 hp to the bit.[4] However, the cutting mechanism of a rotary drilling system using a rolling cone bit requires less energy than other methods of breaking or disintegrating the rock. Diamond bits require somewhat more energy than a rolling cone bit because they crush the rock into smaller fragments.

The low energy required for rock destruction has been one reason conventional rotary drilling has been the almost-universal oil and gas drilling method, despite the fact that much energy is lost in the system between the surface and the bottom of the hole. Still, this lost energy has continued to provide researchers with incentive to study novel drilling methods. Many of these unique approaches to rock destruction can drill faster than conventional drills because they transmit more power to the

Fig. 3-9 Driller at controls of rotary rig. Rotary table with safety guard is at right. (courtesy Oil & Gas Journal*)*

rock; but they also require more energy input. Another factor that has kept these techniques from being put to commercial use is the need for significant modifications of surface equipment.

While these techniques are under study and development, there are still incentives for improving the efficiency of the rotary drilling system, especially in deep holes. Drilling rate generally decreases as well depth increases. One reason is that the pressure of the mud column increases with greater depth, increasing rock strength and inhibiting the removal of rock chips from beneath the bit. Poor removal of cuttings also results in the bit regrinding some rock chips that have already been removed. This uses energy that could otherwise be used to remove new rock chips.

Increasing the power than can be transmitted to the rock is an overall goal of much drilling research. Increasing it within the confines of the power available on a conventional rotary rig is the goal of drilling engineers, drillers, and anyone connected with drilling an individual well.

References
1. Grisham, Jerry. "Hydraulic Rotary Table Built for Apache Rig." *Drilling Contractor*. (February 1982), p. 86.
2. Hook, R.A., L.W. Cooper, and B.R. Payne. "Air Drilling—1: Air or Gas Circulation Can Slash Drilling Time." *Oil & Gas Journal*. (20 June 1977), p. 86.
3. "Task Force Identifies Drill String Techniques." *Drilling Contractor*. (January 1982), p. 40.
4. Maurer, William C. *Novel Drilling Techniques*. Pergamon Press, 1968.

4 Drilling Bits

IT is difficult to single out any one component of the rotary drilling system as the heart of the drilling operation. But the drilling bit is certainly one of the most important items of equipment. It has a significant effect on the cost of drilling the well. Selecting the proper bit for each section of the hole directly influences how fast the hole can be drilled and how severe downhole problems may be.

The number of bits used in a single well varies widely, depending on depth, the type of formations encountered (hard or soft), drilling problems that may occur, and other factors. Under good conditions, the proper bit may be able to drill several hundred feet before having to be replaced, or it may drill only a few feet when a very hard or abrasive formation is encountered.

After the bit is in the hole, it is important that the weight on the bit and the rotating speed for which it was designed be used. Proper bit weight and rotating speed are as important for efficient drilling as is proper bit selection. The correct bit for the formation, if used improperly, will have a shorter life and drilling costs will be higher.

Care is also necessary when lowering the drill string with the bit attached into the hole or removing the bit from the hole. Lack of care during trips in and out of the hole may damage the bit, even resulting in a fishing job if part of the bit is broken off and remains in the hole. If a broken bit must be fished from the hole, drilling cannot proceed until the junk is removed. This may take only hours, but it can also take many days. If a bit is damaged on the trip into the hole and does not drill properly—even though it may not be broken and have to be fished—the time required to complete the extra trip out of the hole and back in with a new bit is costly.

Many operations are required to complete an oil or gas well that do not involve actual drilling. But contractors and operators are always striving to increase the amount of time spent with the bit on bottom. A certain number of trips to change bits is necessary because there is an optimum length of time for drilling with an individual bit. This optimum is determined by considering how fast the bit is drilling as it wears, the bit cost, and the cost of a trip to replace it. But unnecessary trips, fishing jobs, and other interruptions in actual rotating-on-bottom time increase the cost of the well.

Types of bits

Since the advent of the rolling cone bit in the early part of this century, much of the oil and gas well drilling done by rotary rigs has been done with this bit. The general configuration of the three-cone rolling cutter bit has changed little since its introduction. But the development of better cutting surfaces and much more durable bearings has kept pace with the demands of the drilling industry.

The most common rolling cone bit is the steel-tooth or milled-tooth bit, in which the teeth are milled on each cone (Fig. 4–1). The cones are mounted on shafts on the legs of the bit body, offset to increase cutting action, and bearings insure free rotation of the cones on the shafts.

In the early 1950s, the insert bit was developed for drilling hard, abrasive formations. The insert bit has tungsten carbide inserts mounted or inserted in holes in the cone rather than having teeth milled on the cone. The rest of the insert bit configuration is similar to that of the milled-tooth bit—the cones turn on shafts fitted with bearings (Fig. 4–2).

Throughout the development of rolling cone bits, manufacturers have tried to maintain a fairly close balance between expected life of the bearings and expected life of the cutting structure. Either bearing failure or a worn cutting structure means the bit must be replaced. A bearing that lasts significantly longer than the cutting structure is, therefore, of little practical value. And the reverse is also true.

The carbide insert extended the life of the cutting structure. The next key development in the rolling cone bit was the sealed bearing. The seal and lubrication system protected the bearing against the entry of drilling fluid and maintained a clean environment for the bearing.

This combination of a carbide cutting surface and a sealed bearing extended bit life dramatically. Still, the carbide insert bit was applicable primarily in only the harder formations. In the mid-1960s, efforts were made to extend the application of carbide insert bits into medium-hardness formations, but greater bearing life was still needed.

In the late 1960s and early 1970s, the O-ring sealed journal bearing bit

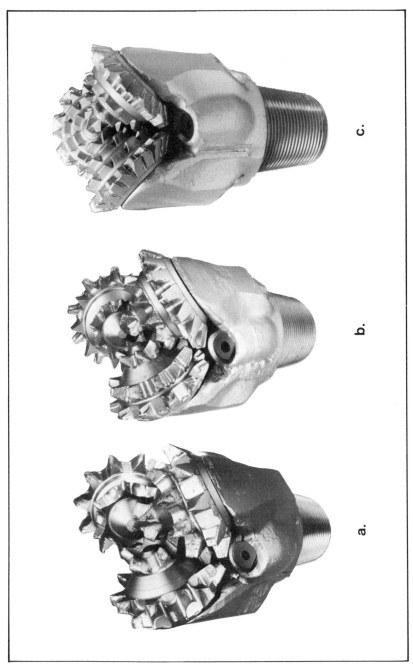

Fig. 4-1 a) Milled-tooth bit for very soft formations, b) Soft to medium formation milled-tooth bit, c) Milled-tooth bit for very high-strength, abrasive formations. (courtesy Hughes Tool Company).

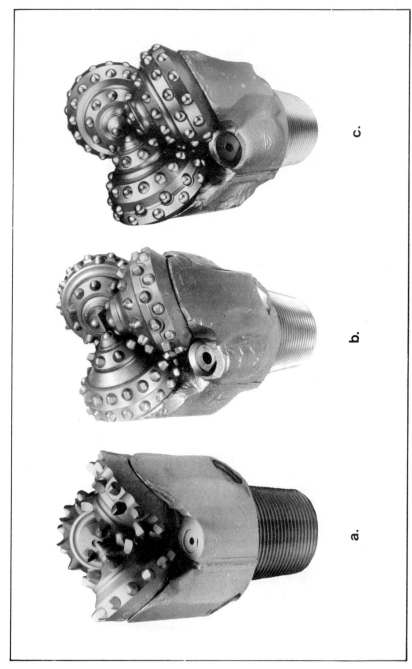

Fig. 4-2 a) Insert bit for very soft formations, b) Recommended insert bit for medium-hard formations with high compressive strength, c) Insert bit for hard, abrasive formations. (courtesy Hughes Tool Company).

was made commercial. This new bearing further increased bearing life and made it possible to design carbide cutting structures for medium and soft formations. Today, carbide bits are applied over a wide range of formation types and hardnesses.

Another adaptation of the rolling cone bit is the two-cone, extended-nozzle bit, designed especially for soft formations. The two-cone carbide insert bit recently developed uses two extended nozzle tubes for better bottom-hole cleaning and a center jet to resist bit balling. In addition to better bit cleaning in soft formations, the design reportedly has shown promise in areas where wells tend to deviate from vertical while drilling—crooked-hole country—because a relatively high penetration rate can be maintained while running at lower bit weights.[1]

In general, lower bit weights tend to maintain a straighter hole. But with most bit designs, penetration rate drops off significantly when the weight on the bit is reduced.

Diamond bits also drill a significant amount of footage in the oil and gas industry (Fig. 4–3). A diamond bit has many small industrial diamonds set in a steel matrix and has no moving parts. Historically used to drill harder formations and in other special situations where very low drilling rates are involved or extra-long bit life is required, diamond bits can also be an alternative to the three-cone rotary bit in many routine drilling operations.

As is the case with many bit designs, the range of application of diamond bits continues to increase. Rather than being a specialized tool for use in very hard formations and with downhole drilling motors, diamond bits are now being used more and more for nonspecialized drilling applications. According to one source, the key to their application, as is the case for most bits, is penetration rate.[2] Conclusions reached from one group of field tests indicate that diamond bits should be considered as an economic alternative to three-cone bits when penetration rates fall below 10–12 ft/hour.

Other types of bits for use with the rotary rig are also available. One relatively new development is the use of polycrystalline diamond compacts in a drag bit designed to drill soft to medium-hard formations (Fig. 4–4). Such a bit is similar in operation to a conventional drag bit in that rock is removed by scraping action. The new bit's teeth protrude from the bit face much more than do the diamonds in a diamond bit, giving it the ability to drill softer formations without becoming clogged. Use of these new bit type is expanding rapidly to a variety of drilling applications.[3]

Other types of bits are under development and have not yet been applied routinely in the field. One example is the core crusher bit. This

74 FUNDAMENTALS OF DRILLING

Fig. 4–3 Diamond bits. (courtesy Petroleum Diamond Products Division, Christensen Diamond Products USA)

diamond bit is designed with a central part like a coring bit. As the bit drills, a core is formed at the center that is broken and then crushed in a chamber located above the bit. The design was aimed at overcoming the fact that in conventional diamond bits a core is formed under certain conditions and removal of cuttings in this core tends to slow penetration rate. Some field testing has been done with the bit and more is planned.

Developments in bit design are aimed at either increasing the life of the bit or making it cut faster. Increasing bit life reduces the number of trips that must be made to replace the bit; drilling faster is an obvious advantage. Both goals of bit development reduce overall drilling cost.

Another approach to reducing the number of trips to replace the bit is

Fig. 4-4 Polycrystalline diamond compact bits. (courtesy Security Division, Dresser Industries, Inc.)

the continuous-chain drilling bit (Fig. 4–5).[4] It was developed as part of the Department of Energy's geothermal well technology program. The objective of the work was to develop and demonstrate in the field a hard rock diamond bit capable of replacing the rock cutting structure downhole.

The continuous-chain bit developed in this program contains 16 cut-

76 FUNDAMENTALS OF DRILLING

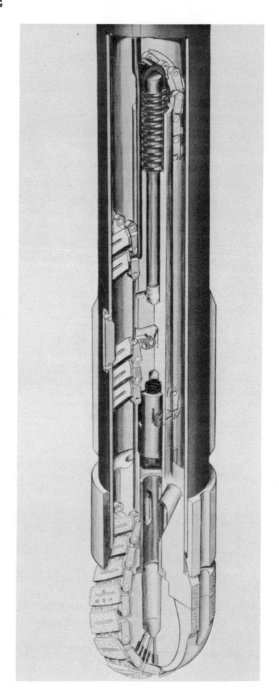

Fig. 4-5 Continuous-chain bit. (courtesy Oil & Gas Journal)

ting surfaces that can be changed without making a trip with the drill string. In the prototype, the cutting structure consisted of both natural and man-made diamonds mounted in a hard matrix. Four polycrystalline diamonds were used to cut the center of the hole, while natural diamonds were used to cut the rest of the hole. The cutting structure was attached to links of a continuous chain that could be cycled downhole by the pressure of the drilling fluid.

The continuous-chain bit advances the hole by using a combination of bit weight and rotary speed as in conventional rotary drilling with other bits. However, it is not yet in commercial oil and gas drilling service.

Other types of bits are also available for special drilling jobs. Coring is a specialized, important part of the analysis of certain wells. It is similar to conventional drilling in some ways, but the purpose of coring is to retrieve a large, undisturbed sample or core of the formation for study.

A special family of coring bits has been developed. These have an opening in the center. A core barrel is mounted above the tool to receive the core of formation cut by the bit. These bits are normally diamond bits. The core cut by the coring apparatus is brought to the surface in the core barrel and is analyzed to determine formation properties and the type and amounts of any fluids in the formation.

General considerations

The drilling bit makes the hole, and anything that can be done to speed the rate at which it removes rock lowers the cost of drilling.

In any well, there is a wide variety of formations to be drilled. Some are very hard; some are relatively much softer. Some are abrasive; some are not. As bit development increased bearing life and cutting structure life, work was also done to design bits to drill a wider range of rock types. Now rotary rock bits are available with several cutting structures designed specifically for formations of all types.

Bits must be replaced when the cutting surface becomes so worn that penetration rate is slowed significantly or when bearings fail and the bit does not rotate properly. Ideally, bearings and cutting surface should fail or wear out at about the same time.

How fast the rock is cut is not the only influence on drilling costs, however. There is an optimum combination of rock-cutting speed and bit durability, or bit life. The search for this optimum is what has kept bit designers busy for decades developing better bearing systems and better cutting structures.

Cost equation. Penetration rate—how fast a bit drills—is a key parameter in overall drilling cost. But it must be considered along with other factors that also affect well cost. For instance, a bit that drills very rapidly but wears out quickly might not be the best choice because it would have to be replaced frequently. Replacing the bit requires pulling the entire drill string out of the hole and lowering it back to bottom with the new bit attached. Drilling expenses—rig rental, labor, fuel—continue while this trip is being made to replace the bit, but no drilling is being done. So it would probably be less costly to use a bit that did not drill quite so fast but was able to maintain an acceptable penetration rate for a much longer period.

This relationship becomes even more important as well depth increases. Not only is more time spent tripping out of the hole to replace the bit with more expense involved, but hole conditions can deteriorate and problems often occur during trips.

Determining the best combinations of penetration rate and bit life is not all guesswork. An equation widely used in the industry to help select the proper bit shows what factors are involved in choosing the right bit:

$$F = [C(L + T) + B]/D$$

where: F = footage cost, \$/ft
C = rig rental rate, \$/hr
L = bit life, hr
T = trip time, hr
B = bit cost, \$
D = depth drilled by the bit, ft

The equation is somewhat empirical since several components are unknown until after the bit has been worn out and replaced. But experience in a well or an area coupled with data on a number of bit runs under similar conditions make the equation very useful in bit selection.

Penetration rate. The design of the cutting structure is an important influence on how fast a bit will drill. In general, when the teeth of the bit—either milled teeth, carbide inserts, or diamonds—protrude only a small distance from the body of the bit or cone, the bit is best suited for relatively hard formations. In softer formations, the teeth must protrude farther from the bit body or cone to provide for adequate cleaning between the teeth. The longer teeth will also remove larger rock chips.

If the teeth are not kept clean, bit balling results or fragments of rock that have been broken off are ground up instead of being swept away by the drilling fluid and carried to the surface. Energy expended to regrind rock that has already been removed is not available to deepen the hole. Thus, inadequate cleaning and regrinding rock fragments reduce penetration rate.

At one end of the formation hardness range, a milled-tooth bit with relatively long teeth would be typical of the bits used to drill soft formations. For extremely hard formations, a diamond bit or a carbide insert bit with relatively short inserts or teeth would likely be used. The medium range of hardness might well be handled by either a milled-tooth, three-cone bit with shorter teeth than those used for the soft formation, a carbide insert bit with inserts designed for medium-hard formations, or, in some cases, a diamond bit. A drag bit with man-made diamond compacts would also be applicable in this hardness range.

The type or shape of the cutting structure is not the only factor that affects penetration rate. In fact, no penetration into the rock is possible unless weight is applied to the bit and the bit is rotated to break up the formation. The amount of bit weight used depends on the type of bit and the type of formation being drilled.

In general, the more weight applied to the bit, the faster it will drill. But there are practical limits on the amount of weight that can be used. Too much weight may damage the bit cutting structure or bearings, causing the bit to have to be replaced prematurely. Too much weight can also cause the bit teeth to be forced into the formation so deeply that they cannot be kept clean.

In some formations, there is also a tendency for the hole to deviate from the vertical. Running too much weight on the bit can aggravate this tendency, resulting in a crooked hole. In fact, when a hole begins to deviate excessively from vertical, one of the remedies is to reduce the weight on the bit and let the pendulum effect of the drill string bring the hole back toward vertical.

Another factor affecting penetration rate is how fast the bit is rotated. The proper rotary speed depends on the type of bit being used and the type of formation being drilled. Rotating the bit too rapidly can damage the bearings and the cutting structure and can shorten bit life.

If bit damage is severe enough to cause the loss of a cone—either as a result of too much weight, too fast a rotating speed, or any other factor—a fishing job may be required to recover the cone before drilling can proceed. If this happens, any gain in penetration rate that was hoped for by running excessive weight or speed is usually much more than offset by the time and expense required for fishing.

Experience has resulted in recommended combinations of weight and rotary speed for many types of bits and formations. Charts have been prepared based on field tests in which penetration rate, bit weight, and rotary speed were recorded. These are a useful guide in well planning. Such tests can also be run while drilling to fine-tune operating conditions for the well being drilled. But there is an almost infinate variety of drilling conditions, and these data can only be used as an estimate. Data for determining the best combination of weight and rotary speed have been quite helpful, however, in drilling more efficiently.

The idea that the best approach is to "turn faster and mash harder" has all but been discarded. True, more weight on the bit increases penetration rate and, in general, a higher rotating speed makes the bit drill faster. But completing a well in the shortest possible time at the least overall cost is not that simple.

Drilling fluid composition and system hydraulics also affect how fast a bit will drill. In general, lighter drilling fluids such as fresh water or brine result in faster penetration rates. The rig must also be able to supply drilling fluid to the bit nozzles at high enough pressures and volumes to keep the bit clean and to carry rock cuttings away from the bottom of the hole and up to the surface.

Rolling cone bits make hole by breaking off chips of the rock as the bit rotates on the hole bottom. How much force is present to hold these chips in place as they are broken from the rock influences drilling rate. A very heavy mud will tend to hold the chips in place more than would be the case with a lighter-weight mud. It is desirable to have the chip removed and carried away from the bit as quickly as possible. If this is not done, the bit must regrind the same chip, and penetration rate will be slowed. So drilling is usually faster with lighter muds.

Adequate hydraulics (fluid circulating conditions) is also necessary to keep the bit clean, especially in softer formations. The bit's nozzles or jets direct the flow of drilling fluid at the bottom of the hole to keep rock cuttings removed from the bit face and from between the teeth.

Information source. Penetration rate is a valuable source of information on what type of formation the bit is drilling. Of course, softer formations will usually drill faster than harder formations, provided the bit is being kept clean and other conditions are comparable.

Significant or abrupt changes in penetration rate may also signal other formation characteristics. For instance, a sharp increase in penetration rate may indicate a more porous formation is being drilled that may be filled with water or hydrocarbons. These fluids are under pressure. If their pressure exceeds the hydrostatic pressure exerted by the drilling fluid in the hole, the chips broken off by the bit will be removed more easily, speeding penetration rate.

Such a rapid increase in penetration rate, often called *drilling breaks*, can be an early warning of an impending kick or blowout. It may indicate that fluid will be entering the wellbore at the bottom of the hole. Other kick-detection methods should then be checked closely.

A significant decrease in penetration rate may indicate bit damage other than normal wear. But a drop in drilling rate can also indicate a change in the type of formation being drilled or an adverse change in some other drilling factor.

Bit selection

Much effort goes into bit selection during well planning. Modifications will, however, be made to the plan as the well is being drilled. Unexpected situations occur that call for a special bit or a change in the plan for a particular interval in the well.

Unless the well is a rank wildcat, far from any other wells, a considerable amount of information is available on which to base a well plan, including bit selection. Logs of nearby wells and bit records and penetration rate records made during the drilling of those wells give a very good indication of the types of formations that will be encountered. Even in the case of a wildcat, the geologist will prepare an outline of the expected formations, and information will be available on at least the general physical properties of the formations.

A bit record contains the type of bit used, how long it was used, and other information. When these data are combined with penetration rate data from the well record, it is possible to estimate closely which type of bit will drill the fastest in each of the expected formations.

In addition to these records, bit manufacturers have much information concerning how each type of bit has performed in a given area under specific conditions. This information is available when selecting the proper bit during well planning and drilling. Manufacturers also have computer programs that analyze rig capability relative to certain types of bits and other programs that compare the economics of different types of bits under similar drilling conditions.

As mentioned, the type of rig used to drill the well must also be considered in bit selection. The rig must have sufficient pump capacity, for instance, to drill with certain types of bits. This may be particularly important when drilling with a downhole drilling motor, which depends on adequate drilling fluid circulation for its power.

Example weights and speeds. It is impossible to say how much weight should be used and how fast the bit should be rotated in a specific formation. Even though bit manufacturers provide general guidelines developed from experience, both of these parameters must be adjusted during an individual bit run and will be different in each well.

Since a given bit type (a milled-tooth bit designed for very soft formations, for example) is available in several sizes, the recommended weight on bit is often given in pounds per inch of bit diameter. Then the driller calculates the total weight on the bit he should use, depending on the size of the bit being run.

In general, the recommended weight on bit for softer formations is less than that for harder formations. For example, the recommended weight to be run on a typical milled-tooth bit for very soft formations is 3,000–5,000 lb/in. of diameter, while a typical bit for a very high-strength, abrasive formation should be run with 6,000–8,000 lb/in. of diameter.

The same trend is true for insert bits. Recommended bit weight for a typical soft-formation insert bit is 2,500–4,500 lb/in. of diameter, while the recommended weight on bit for the insert bit used in hard formations is 4,500–6,000 lb/in. of bit diameter.

Rotary speeds recommended by manufacturers often decrease as the formation hardness increases. It is usually recommended that, within the recommended rotary speed, the lower speeds be used with higher weights on bit. For example, the manufacturer recommends the bit shown in Fig. 4–1a be run at 120–90 rpm, the bit in Fig. 4–1b at 100–60 rpm, and the very hard formation bit in Fig. 4–1c at 70–50 rpm. Recommended rotary speeds for the insert bit in Fig. 4–2a are 150–60 rpm, while the hard-formation bit in Fig. 4–2c should be run at 60–45 rpm.

References
1. Baker, William. "Extended Nozzle Bits Require Precise Nozzle Sizing." *Oil & Gas Journal*. (19 March 1979), p. 88.
2. Striegler, John H. "Recent Field Applications Show Diamond Bits Economical." *Oil & Gas Journal*. (23 July 1979), p. 51.
3. Slack, James B., and Jeffrey E. Wood. "Stratapax Bits Prove Economical in Austin Chalk." *Oil & Gas Journal*. (24 August 1981), p. 164.
4. "Core System, Chain Bit Get Field Test." *Oil & Gas Journal*. (8 October 1979), p. 188.

5 Drilling Fluids

THE key to making the rotary drilling system work is the ability to circulate a fluid continuously down through the drill pipe, out through the bit nozzles, and back to the surface. Though often thought of as a liquid, the drilling fluid can be air, foam (a combination of air and liquid), or a liquid. The engineering definition of *fluid* includes both gases and liquids, so fluid is properly used to describe all circulating media. Liquid drilling fluids are commonly called *drilling mud*.

All drilling fluids, especially drilling mud, can have a wide range of chemical and physical properties. These properties are specifically designed for drilling conditions and the special problems that must be handled in drilling a given well. There are, however, a number of general types of drilling muds that are widely used, and their properties are adjusted to fit conditions in individual wells.

The worldwide market for drilling fluids, drilling fluid handling equipment, and related technical services was estimated to be $2.4 billion in 1980. Revenues were expected to jump to $3.25 billion in 1981.[1] As a percentage of the cost of drilling a well in the U.S., drilling fluids accounted for about 7% in 1980.

Circulation path

The drilling fluid flows through a number of equipment components in going from the surface to the bottom of the hole and back to surface.

When liquids are used as a drilling fluid, large positive-displacement pumps are the starting point in the circulation path. From the pump discharge, fluid flows through piping to the standpipe on the rig floor. A rotary hose (Fig. 5–1) connected to the standpipe provides a flexible connection that is needed as the drill pipe is lowered into the hole while drilling. The rotary hose connects to the drilling swivel, which provides

Fig. 5-1 Rotary hose connects stand pipe and swivel.

the connection for the flow path between the stationary hose and the rotating drill string. Mud flows down through the center of the drill string, drill collars, and any other tools in the drill string, then out the nozzles or jets in the bit.

After picking up cuttings removed by the bit, the drilling fluid moves up the hole through the annulus, the space between the outside of the drill pipe and the wall of the hole or the casing. At the surface, mud flows to the shale shakers where larger cuttings are removed. Then mud may pass through other solids-removal equipment, the type and number of units depending on well conditions. After flowing through a settling tank, it goes to the pump suction tank to be pumped downhole again.

Mixing equipment is also included in the surface mud system for adding materials to the mud stream to adjust its properties.

These equipment items are common to most rig mud systems handling liquid drilling fluids. When air or foam is used as the drilling fluid, additional equipment is needed. The main additional item required for air drilling is one or more compressors to force air down the drill string and up through the annulus. Liquid muds are pumped downhole, but air must be compressed.

Air compressors of the capacity needed for air or foam drilling are not standard equipment on most rotary drilling rigs, so another firm is normally called on to provide air drilling equipment and service. The compressor package is usually truck-mounted, making it compact and highly mobile.

When drilling mud replaces air drilling or the well is completed, the air compression equipment is released from the location.

Purpose of drilling fluids

The drilling fluid serves several functions. It lubricates and cools the bit as it breaks up the rock at the bottom of the hole. It carries the rock cuttings to the surface where they are removed from the drilling fluid before it is recirculated. It helps control pressures that exist in formations penetrated by the bit. And it is a valuable source of downhole information.

Cooling and lubrication. As the bit drills into the rock formation, the friction caused by the rotating bit being forced against the rock generates heat, much as a twist drill used in a home workshop does when drilling in wood or metal. If this heat is not dissipated, the bit cutting structure, bearings, and other components become damaged and bit life is shortened. The drilling fluid absorbs much of this heat as it is circulated past the area where the bit contacts the rock.

In addition to the heat generated by friction between the bit and the rock, the static downhole temperature—the temperature of the rock excluding the heat generated by friction and the cooling effect of the drilling fluid—increases with depth. This increase in formation temperature with depth is termed the *temperature gradient* and is usually assumed to be a 1°F increase for each 100 ft of depth.

The temperature gradient is added to surface temperature to estimate the bottom-hole temperature at any depth. For instance, assuming a 50°F surface temperature, the bottom-hole temperature in a well 15,000 ft deep would be estimated as:

$$(15,000/100)(1) + 50 = 200°F$$

Subsurface temperature gradients vary from this 1°F/100 ft norm in different areas of the world. Although a given gradient exists over a long interval of hole, there may be variations from that gradient over short distances within that interval.

In addition to serving as a cooling medium, the drilling fluid lubricates the bit as it contacts the rock at the bottom of the hole. Since most rock bits remove rock by chipping or gouging rather than grinding, the lubrication effect does not significantly affect how fast the bit drills. Drilling mud also provides some lubrication for the drill string throughout the hole where it contacts the walls of the hole or the casing, reducing wear on the drill pipe, tool joints, and other components.

Cuttings removal. An important function of the drilling fluid, whether air, gas, or foam, is to carry rock cuttings removed by the bit to the surface. There the drilling fluid flows through treating equipment where the cuttings are removed and the clean fluid is again pumped down through the drill pipe string.

These cuttings or drilled solids must be removed from the hole so the bit can continue its advance into virgin rock. Drilled solids must be removed from the drilling fluid so its desired physical and chemical properties are maintained. Often, special solid materials are added to the drilling mud at the surface to give it a desired property. If drilled solids are allowed to build up in the fluid without being removed, the effect of these added solids is changed. Likewise, the drilled solids must be removed from beneath and around the bit as quickly as possible so the bit does not expend energy redrilling the same rock.

The drilling fluid must be designed to carry the drilled solids to the surface with a minimum of settling. Many other considerations are also necessary in formulating the proper drilling fluid for expected hole conditions, but handling drilled solids may be its most important function.

Pressure control. The drilling mud can be the first line of defense against a blowout or loss of well control caused by formation pressures. After the drilling mud is designed for expected conditions, its properties must often be adjusted during drilling to provide well control. Making the correct adjustments requires keeping constant watch on drilling conditions, cuttings, and other data.

The hydrostatic head, or pressure, that the drilling fluid exerts on the wall of the hole is the property that controls formation pressure. This pressure may need to be adjusted several times during the drilling of a single well as formations with different pressures are encountered.

Hydrostatic pressure of the drilling mud is calculated using the weight of the mud in pounds per gallon, the depth of the formation, and a conversion factor that converts mud weight in pounds per gallon to pressure in pounds per square inch. For example, the hydrostatic pressure in pounds per square inch is equal to:

$$(\text{depth}) \times (\text{mud weight}) \times (0.052)$$

Fresh water weighs 8.34 lb/gal and is often used to drill portions of the hole where abnormal formation pressures are not expected and where fresh water will not cause the wall of the hole to become unstable. As an example of the calculation of hydrostatic pressure, assume a rig is drilling at 10,000 ft with fresh water. The hydrostatic pressure of the drilling fluid at the bottom of the hole would be:

$$(10,000) \times (8.34) \times (0.052) = 4,337 \text{ psi}$$

where 0.052 is the factor to convert pounds per gallon to pounds per square inch.

Just as there are normal temperature gradients, there are also normal pressure gradients. Accepted definitions of normal pressure gradients are 0.465 psi/ft of depth in salt-water basins (geologic basins in which the pore spaces of the rock contain salt water) and 0.429 psi/ft of depth in brackish or fresh-water basins. This means that formation pressures would be expected to increase by 0.465 psi (in the case of a salt-water basin) for each foot of depth.

Gradients that exceed these values (where pressure increases at a faster rate with depth) are termed abnormal pressure gradients, or geopressures. Subnormal pressure gradients also exist where the increase in formation pressure with depth is less than these values.

If the well in the example above were being drilled in a salt-water basin, the pressure in the formation at 10,000 ft would be expected to equal:

$$(10,000) \times (0.465) = 4,650 \text{ psi}$$

If this were the case and fluids existed in the formation at 10,000 ft, a fresh-water drilling fluid would not be adequate to contain the reservoir pressure. A slightly heavier drilling fluid—heavier than 8.34 lb/gal—would have to be used.

It is not enough to design a drilling fluid that has sufficient weight to contain any pressures that could possibly be encountered by the bit. Though one formation may contain fluids under pressure that must be controlled by a relatively heavy mud, a hole interval above the high-pressure section may not be able to withstand the hydrostatic pressure of the heavier mud.

If the mud weight is increased to control the high-pressure formation, the weaker formation above may break down under the hydrostatic mud pressure. When this happens, drilling fluid will flow into the weaker formation, a condition known as *loss of circulation*.

Lost circulation occurs to varying degrees, and the loss of a portion of the drilling fluid into a formation can often be tolerated, depending on the situation. But the greater the volume of drilling fluid lost, the more difficult well control becomes. It also becomes more difficult and expensive to maintain the proper volume and properties of the drilling fluid because large additions must constantly be made at the surface. If the drilling fluid must contain expensive additives to combat difficult hole problems, then lost circulation is especially costly.

When a large portion of the mud being pumped into the well flows into a formation, little—maybe none—returns to the surface. This means an important source of drilling information is not available.

There are several solutions to lost circulation problems. The first attempt to stop lost circulation may be made by adding lost circulation materials (LCM) to the drilling mud at the surface. As the mud is circulated, the LCM bridges, or plugs, the spaces in the formation that are taking fluid.

In the most severe cases, however, it may be necessary to install casing to a point below the weaker formation. With the weaker formation behind pipe, it is possible to increase the weight of the drilling mud to that necessary to control the high-pressure zones below. The steel casing will easily contain the higher hydrostatic pressure through the weaker zone.

These two zones—one relatively weak into which mud will flow and the other containing relatively high pressure—may be close together. A thin, impermeable rock layer may separate the two zones. In this case, it is critical that the depth at which the casing will be set be chosen accurately. Often, the impermeable layer separating the two formations is only a few feet thick and the bottom of the casing must be set within that few feet to isolate the weaker zone before entering the high-pressure zone.

Analysis of cuttings and other drilling factors is the key to selecting this depth. If drilling proceeds into the high-pressure formation before casing is installed, serious problems can result.

The drilling fluid also is used to solve or to avoid other problems that may slow drilling, cause fishing jobs to be required, or eventually cause loss of either part or all of the hole. There is a wide variety of drilling mud additives and mud formulations available to solve specific drilling problems, including hole sloughing, swelling formations, leaching in salt sections, and corrosion.

Data source. The drilling fluid is a source of information for the geologist, the drilling engineer, and the rig crew. The cuttings it brings to the surface can tell the geologist the type of formation being drilled. Changes in the volume of drilling fluid returning to the surface can warn of dangerous hole conditions. Cuttings removed from the hole by the drilling fluid serve an important function during drilling. The geologist on the well monitors these cuttings carefully and continuously to analyze the type of formation being drilled. By examining rock fragments brought to the surface by the drilling fluid, he can tell what type of rock is being drilled. He can also compare what the cuttings tell him with what had been expected at that depth when the well was planned, and he can more accurately project what types of formations lie ahead.

For instance, if he had expected to see layers of different types of rock in a certain order in the hole and the first few layers penetrated by the bit are in that expected order but at a shallower depth, then he will expect the remaining layers to occur at shallower depths than originally estimated. Both the identification of formation types being drilled and the projection of the depths of other formations are keys to determining at what depth to install a casing string, for example.

As much advance warning as possible of the depth at which each formation will occur is necessary, particularly in the case of exploration wells. The lack of adequate field data prior to drilling such wells makes analysis of cuttings during drilling more important. For example, it is critical in many cases that the casing point, the depth at which the bottom of a string of casing will be set, is selected within a tolerance of a few feet. Knowing when this casing must be installed is necessary in planning casing cementing and other operations.

To make the information provided by the cuttings accurate and useful, the geologist must consider *lag*. Lag is the time it takes a rock fragment that is removed by the bit to reach the surface. When the geologist is able to view a rock fragment that the mud has brought to the surface, the bit has already drilled an additional number of feet. Unless drilling was halted immediately following the removal of a particular rock cutting, the cutting was removed from a shallower depth than that at which the bit is drilling when the cutting reaches the surface. So to calculate the depth from which the cutting being analyzed was removed, it is necessary to include lag.

Also called bottoms-up time, lag is determined by dividing the volume of the annular space (volume between the drill pipe and the walls of the hole or the casing) by the mud pump output. If the annular volume is known in barrels and the pump output is in barrels per minute, the lag time calculated will be in minutes.

Drilling fluid can solve problems

Many drilling problems are the result of conditions or situations that occur after drilling begins and for which the drilling fluid was not designed. Some of these problems can be solved by adding materials to the drilling fluid to adjust its properties. In other cases, it may be necessary to replace the drilling fluid being used with another fluid system.

One of the most common changes made in a drilling mud is its weight, or density. Weighting material is added when high-pressure formations are expected. If a long section is to be drilled in which formation pressures are not high, it may be desirable to reduce the mud weight. In general, a lighter mud lets the bit drill faster.

Besides changes in weight, adjustments are made in the drilling fluid to handle other problems.

Lost circulation. One common drilling problem is the loss of drilling fluid into a zone in the well bore. Lost circulation can occur in several types of formations, including highly permeable formations, fractured formations, and cavernous zones containing large voids or channels.[2] When this happens, the fluid column loses at least some of its ability to control well bore pressures and perform its other functions.

A key factor in preventing loss of circulation is to keep a close watch on mud density to keep the hydrostatic mud pressure as low as possible yet high enough to control formation pressures in other zones. Another important consideration is the depth at which casing strings are installed. If casing is set too shallow, formations below the bottom of the casing may not be strong enough to support the hydrostatic mud pressure needed to control formation pressures lower in the hole.

Lost circulation may occur while using almost any fluid, but weighted drilling muds are more likely to cause fluid loss. Several methods exist for handling the problem, depending on well conditions. But there are instances when formations are so weak they will not even support the hydrostatic head exerted by fresh water, which does not contain any weight material. In these wells, air or foam can be used as the drilling fluid, providing a much lighter column of fluid that can be supported by the weak formations.

In liquid mud systems, lost circulation materials can be added to the mud to bridge or deposit a mat where the drilling fluid is being lost to the formation. These materials include cane and wood fibers, cellophane flakes, and nut hulls. The type of material depends on, among other factors, the size of the crevices or fractures into which drilling mud is being lost.

Another approach to sealing a lost circulation zone is the use of plugs, a precise volume of a material that is circulated to the lost circulation zone to seal it off.

Changes in drilling mud properties and drilling practices, in addition to the use of special materials, are also made while drilling to help reduce lost circulation. Some experts recommend the following steps be taken if lost circulation occurs, depending on conditions:

1. reduce circulation rate
2. reduce mud weight
3. adjust other properties of the mud besides its weight
4. examine drilling conditions closely to determine if the pipe is being pulled or lowered into the hole too fast on trips, if balling of the bit is occurring, or if changes need to be made in the bottom-hole assembly.

Stuck pipe. The drilling fluid can also be designed or its properties can be adjusted while drilling to prevent another common drilling problem: stuck pipe. Many drilling fluids form a filter cake, or layer of wet mud solids, on the wall of the hole in permeable formations. The hydrostatic pressure of the mud column can then press the drill string into this filter cake against the wall of the hole in formations where the pressure is lower than the hydrostatic pressure of the drilling mud. Solids are deposited around the drill pipe and the pipe becomes stuck.

This can occur after drilling has been halted for a rig breakdown, while running a directional survey, or when conducting other nondrilling operations. If the condition is severe, it may be that the time required to add another joint of pipe at the surface in order to continue drilling is sufficient for the pipe to become stuck.

Several preventive measures during drilling can be taken to avoid stuck pipe:

1. keep the pipe moving vertically or rotating
2. avoid shutdowns of long periods, if possible
3. speed the rate at which connections are made

Also, the use of grooved drill collars reduces the tendency for the pipe to become stuck since the area in contact with the hole wall is reduced and circulation paths are present to equalize pressure and remove solids.

In addition to these mechanical changes in drilling operations, the drilling mud can reduce the tendency of the pipe to stick. Using a mud weight that is as low as possible, considering formation pressures, adjusting mud properties to provide a firmer filter cake, and adding an emulsifier to the mud can all reduce pipe sticking.

Spotting oil around the pipe at the point at which it is stuck and allowing the oil to soak is a common method of freeing struck pipe. This is done by pumping a volume of oil around—down through the drill pipe and up the annulus to where the pipe is stuck—with the drilling fluid.

To spot a material effectively, it is necessary first to know where the pipe is stuck—to find the free point. The free point is the point above which the pipe is free. It may be found using a special tool, or it may be estimated using a formula based on the stretch of the drill pipe.

By pulling on the stuck pipe and raising the surface end of the drill string, data can be obtained to use in the formula. The amount of force needed to raise the surface end of the drill string and the distance it was raised are measured. The weight of the pipe is known, and the free point can then be estimated.

Heaving or sloughing shale. Other hole problems that can be reduced or eliminated by proper mud design include heaving or sloughing shale. This occurs when shales enter the well bore after the section has been penetrated by the bit. Some solutions to the problem are mechanical: improving mud hydraulics by changing pumping equipment or bit nozzle size, changing the amount of weight imposed on the bit, using more drill collars to make the drill string stiffer, and insuring the hole is kept full of fluid at all times.

Often the solution lies in an adjustment of drilling mud properties. Increasing mud weight to hold the formation in place or raising the viscosity of the mud so cuttings will be carried to the surface more efficiently are common steps taken.

If such problems occur, it may be necessary to suspend drilling for a period and condition the hole. The drill string is left in the hole, although the bit may be raised above the bottom of the hole, and drilling mud is circulated for a period to improve hole stability. Adjustments may be made in drilling fluid properties before the conditioning period if hole problems exist. Little change may be necessary, however, prior to conditioning the hole in preparation for installing casing.

Types of drilling fluids

Of the three general types of drilling fluids—water base, oil base, and air/foam—water-base fluids are used most often. In general, the components of a water-base system and the equipment used to circulate and treat it are more common than for the other two types. Additional expense is often involved, for instance, when drilling with air because the equipment needed is not part of the standard rotary rig. Nevertheless, all three systems have application, and any one may be required to handle specific conditions.

More than one of these types of fluids may be used on a single well. Drilling conditions may change unexpectedly, requiring properties that cannot be provided by one type of fluid. Or the drilling plan, because of experience in nearby wells or other available data, may call for changing from one type of fluid to another at a specific depth.

Water-base muds. Water-base drilling fluids are muds in which water is the continuous phase.[3] The continuous phase is that which contacts the formation and the well bore. A discontinuous phase may or may not be present. If it is (it might be oil added to the mud, for example), the discontinuous phase is segregated into droplets that do not contact any surface except that of water.

Water-base drilling fluids, in addition to various muds to which have been added a variety of materials, include fresh-water and low-solids brines. A number of salts such as sodium chloride, calcium chloride, or potassium chloride may be added to fresh water to achieve desired properties. Salts can increase the weight of a drilling mud for pressure control purposes without adding solids to the mud. And salts often help reduce formation damage and hole problems in clay and other formations. For instance, they can inhibit swelling in clays.

Sea water is often used in offshore drilling, primarily because of its availability rather than its particular properties.

A wide variety of water-base mud systems is available. By using a combination of the many additives offered by mud suppliers, almost any desired property can be obtained.

The design of mud systems is complex and cannot be detailed here. It involves both chemical and physical properties of the formations to be drilled and the fluid to be used. Designing the precise system for best results is a highly specialized science.

In addition to fresh-water, salt-water, and sea-water muds, others include lignosulfonate muds, which offer inhibition and are compatible with fresh-water, salt-water, and calcium muds. They are often used in deep drilling where high temperatures and pressures exist. Lime muds are often used where sloughing shales cause problems during drilling, and gyp muds are used in troublesome shale formations in addition to being used in thick sections of anhydrite or gypsum. Saturated salt-water muds, which are used primarily when drilling through very thick salt beds or salt domes, can prevent excessive hole enlargement resulting from leaching the salt section. Nondispersed muds can increase drilling rate, reduce lost circulation, and minimize wear on bits and pumping equipment.

Oil-base muds. An oil-base mud is a drilling mud in which oil is the continuous phase. These muds, like water-base muds, are used in drilling, completion, and workover operations. They are used where free water is not desirable or where special properties are needed to handle a specific problem.

Some common applications of oil-base drilling fluids include drilling through formations that are sensitive to water, drilling through zones where hole enlargement or sloughing shale is a problem, and drilling where conditions that cause differential sticking of the drill pipe are expected. As mentioned previously, an oil-base fluid may also be used to free the drill string after it has become stuck.

Oil-base fluids are also used on wells in which high temperatures exist. In deep holes, bottom-hole temperatures may be such that water-base muds will not maintain their desired properties. And they can be used in corrosive environments that can damage drill pipe and other equipment.

Diesel oil is widely used to provide the oil phase of oil-base muds because it is readily available and its properties are uniform.

In addition to the oil phase, a wide range of additives can be included in oil-base muds to provide desired properties, just as is the case with water-base muds. The water phase of the fluid often contains a salt. Other materials are added to improve stability in high temperatures, to provide better solids suspension ability, and to increase the weight of the fluid to control formation pressures. And the weight of oil-base muds can be increased for formation-pressure control, just as is the case with water-base muds.

Special rig equipment is often recommended when using oil-base muds. Much of it is aimed at preventing loss of the mud when pulling the drill string from the hole. Since diesel oil is more costly than water used in a water-base fluid, preventing unnecessary loss of an oil-base mud is especially important.

One consideration in the use of oil muds is that their treatment and the disposal of cuttings from oil-mud systems may require special equipment and techniques to meet environmental regulations, especially in the case of offshore drilling. Approaches to this problem have been approved by various authorities.

Air and foam. There are drilling conditions under which a liquid drilling fluid is not the most desirable circulating medium. Air or foam is used in drilling some wells when these special conditions exist.

In general, air drilling is used where significant volumes of formation liquids or high pressures are not expected. Air drilling may be used to drill a portion of an individual well; then the drilling fluid can be converted to a liquid system.

Air can perform some of the functions required of a drilling fluid. It can absorb heat generated by friction resulting from the bit's contact with the formation, and it can move the rock chips removed by the bit away from the bottom of the hole. In fact, since the hydrostatic pressure of a column of air is much less than that of a column of liquid in the hole, rock fragments are more easily removed from the formation because air exerts less holddown force on the rock face. Air drilling, therefore, results in increased penetration rates compared with those possible under the same conditions when a liquid is used as the drilling fluid.

But an air drilling fluid system does not have the cuttings-carrying capacity of a liquid; and because of its light weight, it cannot control formation pressures as well as a liquid mud system can.

Air drilling, where only air is used as the circulating medium, has been modified to obtain certain properties by adding water and other materials to the air stream to produce a mist or foam. Two types of foam drilling fluids have been developed: stiff foam and stable foam.[4] Stiff foam was developed for drilling very large holes for purposes other than oil and gas production. The large diameter of these holes, some of which were drilled by the U.S. Atomic Energy Commission, made it difficult to carry the cuttings to the surface with air. So a gel-base mud was designed for this purpose. But this type foam is not used in oil and gas drilling operations.

A stable foam, a completely mixed combination of air and liquid in which the liquid is the continuous phase, has been used for drilling oil and gas wells. Such fluid systems must be carefully designed and maintained to provide the desired characteristics.

In addition to the higher penetration rates possible when using air as the drilling fluid, there is little cost for chemical additives and water requirements are low. However, there are disadvantages. If a water flow occurs in a zone that is penetrated by the bit, drilling problems can result. Or if air is used to drill through formations that are not well consolidated, erosion of the wall of the hole can occur. And if hydrocarbons are encountered while drilling, the possibility of a downhole fire exists.

Mud properties

Many years of research and development work have led to a thorough understanding of how drilling fluid properties affect penetration rate, hole stability, and equipment. With this knowledge, a wide array of sophisticated drilling fluid systems has been designed to cope with a variety of problems.

One of the most important concerns while drilling is solids control. This is accomplished by changing the properties of the drilling fluid and by using mechanical equipment at the surface to remove solids from the circulating mud stream. When the amount of solids in the mud is not controlled, the rate of drilling is reduced and the possibility of other drilling problems such as lost circulation is increased. Both lower penetration rates and problems caused by excessive solids in the mud increase the cost of the well.

There are two sources of mud solids. Some solids are added at the surface to increase the density of the fluid (weighting material) or to change other properties of the mud. A few of these solids added to the mud at the surface are active; others such as barite are inert. The other source of solids is the formation being drilled.

As discussed earlier, the weight of a drilling fluid can be adjusted over a wide range in order to control pressures in zones penetrated by the bit. Drilling fluid density can range from that of fresh water, 8.34 lb/gal, to very heavy fluids weighing more than 20 lb/gal. The ideal conditions exist when the lowest density fluid is used. But to control wellbore pressures, it is often necessary to weight up the drilling mud.

The most common material used to increase mud density is barite (barium sulfate). Its specific gravity is 4.2–4.3. By comparison, water has a specific gravity of 1.0. But other materials besides barite can be used to increase mud density. Recent drilling using ground ilmenite as weight material indicated that the use of this heavier substance (specific gravity of 4.58) could increase penetration rate.[5]

Since the solids content by volume of a drilling mud is a critical factor in drilling efficiency, especially in the case of heavy muds, the higher density weight material provides a given mud weight with less volume of weight material. If the maximum volume of solids that can be tolerated in a 20-lb/gal drilling mud is 45%, for example, the barite needed to raise the density to 20 lb/gal would make up more than 43% of the fluid by volume. This would leave little room for adding other solids that might be needed to give other properties to the drilling fluid. If the heavier ilmenite were used instead of barite to form a 20-lb/gal mud, the ilmenite would comprise only 39% by volume of the drilling fluid. Then 6% of the mud volume would be available for other needed solids without exceeding the 45%-by-volume limits.

Of course, density is only one property of the drilling fluid that is adjusted to make the mud perform its several duties in the most efficient way. Other materials are added to the drilling fluid, and it may be necessary to change mud treatment to remove excessive solids.

The mud must be monitored constantly to detect changes that may reduce drilling rates or cause other hole problems. Some of the key drilling fluid properties that are monitored in addition to density are fluid loss, viscosity, yield point, gel strength, and filtrate.

When a drilling fluid loses water to a permeable zone, the solids in the drilling fluid build up on the wall of the drilled hole.[6] Excessive buildup can cause stuck pipe and other problems.

A close watch on the *viscosity* of a drilling fluid can warn of potential problems resulting from solids in the mud.[7] Also, as plastic viscosity—one measurement made on the mud stream—increases, the pressure drop in the mud circulating system increases.

Yield point is influenced by the concentration of solids, their electrical charge, and other factors. If not at the proper value, it can also reduce drilling efficiency by cutting penetration rate, increasing circulating pressure, and posing the danger of lost circulation.

While proper *gel strength* can help suspend solids in the hole and allow them to settle out on the surface, excessive gel strength can cause a number of drilling problems.

The rate at which *filtrate* will invade a permeable zone and the thickness of the filter cake that will be deposited on the wall of the hole as filtration takes place are important keys to trouble-free drilling.

Drilling fluid treating and monitoring equipment

In addition to the main mud pumps, several items of mud treating equipment are found on most rigs (Fig 5-2). Much of this equipment is aimed at solids removal, including shale shakers, desanders, desilters, and centrifuges. Although the specific items used for a given mud system will differ—depending, for instance, on whether or not a weighted mud system is being used—this equipment is included in many mud-handling systems.

Shale shakers (Fig. 5-3) remove the larger particles from the mud stream as it returns from the bottom of the hole. Shakers can be equipped with screens of various sizes, depending on the type of solids to be removed. More than one shaker may be used on a single well with more than one screen size.

Finer particles in the mud stream are removed with desanders, desilters, and centrifuges (Fig. 5-4). Each of these items of solids-control equipment is applicable only over a certain range of particle sizes. It is important that this mud treating equipment be installed properly so the mud stream proceeds through the different types of equipment in the proper sequence. Otherwise, efficient solids removal is not possible.

In addition to removing solids, mud handling equipment may also include a mud degasser (Fig. 5-5) to remove entrained gas from the mud stream. Degassing the drilling fluid is sometimes necessary when small volumes of gas flow into the well bore during drilling. These volumes are not large enough to pose the danger of a blowout. But if not removed, the gas will reduce the density of the mud and additional gas may enter the hole. Gas in the drilling fluid also adversely affects operation of mud pumping equipment.

The tendency of entrained gas bubbles to remain in the mud varies with the type of drilling fluid.[8] Gas bubbles break out of water or brine easily, and a degasser may not be needed when drilling with these fluids. Other muds with higher viscosities tend to hold entrained gas bubbles, so a degasser may be required.

Most degassers work by getting the gas bubbles to the surface of the liquid so they can burst. In some degassers, mud flows across a steel plate in a thin layer. The gas bubbles rise to the surface and break out. In other units, the mud is sprayed in a thin sheet against a wall, causing the gas bubbles to burst.

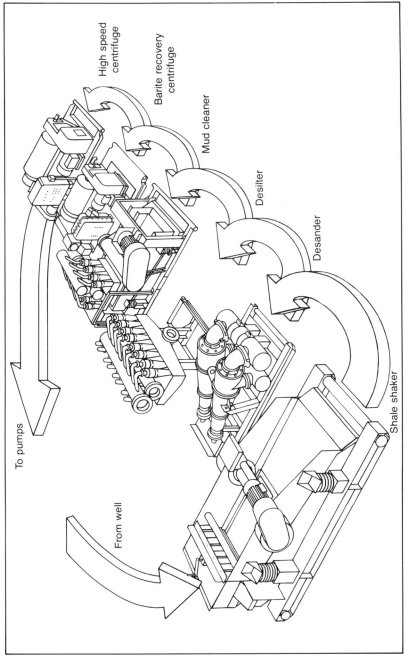

Fig. 5-2 Example mud system layout. (courtesy Swaco Division, Dresser Industries, Inc.)

100 FUNDAMENTALS OF DRILLING

Fig. 5-3 Shale shaker removes large particles from mud stream. (courtesy Swaco Division, Dresser Industries, Inc.)

Additional equipment is also part of most mud systems, including mixers to agitate mud in the tanks, smaller pumps for various duties, and equipment for adding chemicals and solid materials to the mud system.

Air drilling calls for special equipment, as discussed earlier (Fig. 5–6).

Advances in drilling fluids

Despite the range of drilling mud systems and additives available today, development is underway on new mud additives and on better application of today's fluids. To keep drilling costs in line, there will probably be continued emphasis on solids control.

One expert says the future will see increased use of oil-base muds.[9] He sees the industry taking more advantage of the benefits that oil-base muds can offer. One benefit is well bore stabilization and lubrication. An important application may be in directional wells drilled at high

DRILLING FLUIDS 101

Fig. 5-4 a) Vertical desander, b) Desilter, c) Centrifuge are equipment items for removing finer particles from mud. (courtesy Swaco Division, Dresser Industries, Inc.)

102 FUNDAMENTALS OF DRILLING

Fig. 5–5 Mud degasser. (courtesy Swaco Division, Dresser Industries Inc.)

angles for long distances. Oil muds can lower friction between the drill string and the hole while rotating and while tripping in and out of the hole. A second benefit is in temperature stability and corrosion protection. Since wellbore temperatures increase with depth, growth in deep well drilling would expand the use of oil muds. Oil muds are among the most thermally stable drilling fluids available. Finally, oil-base muds can help prevent formation damage. After drilling is complete, the effects of oil-base muds on possible producing formations are relatively easily removed by the produced fluid.

In the case of water-base muds, efforts are underway to expand the temperature range in which they can be used. New additives will be capable of meeting environmental regulations. There may also be substitutes developed for barite, both because of advantages offered by other weighting materials and to conserve barite reserves.

All of this work will be done because of increasing drilling costs. If the industry is going to drill the number of wells required to find oil and gas in the amounts needed to satisfy world demand, it will be necessary to cut costs wherever possible and to speed the drilling operation.

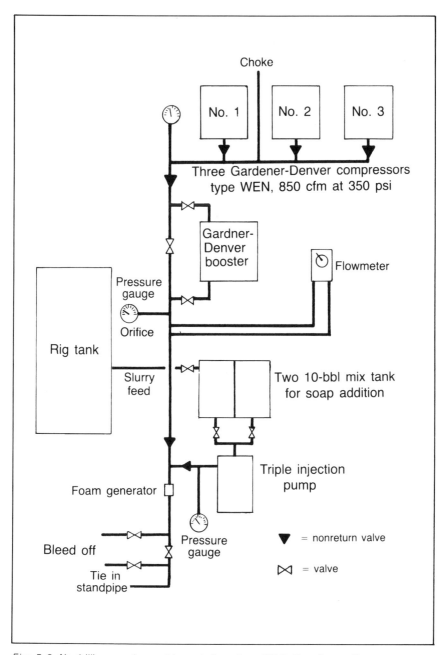

Fig. 5-6 Air drilling equipment layout. (courtesy Oil & Gas Journal)

TABLE 5-1
Per barrel cost of common drilling fluids

Weight, lb/gal	Water-base fluids			Oil-base fluids	
	Benex nondispersed	Lignosulfonate	X-C Drispac 3% KCl	Invert HFL	Oil mud
9	$4.00	$9.00	$21.00	$76.00	$80.00
12	19.00	24.00	37.00	85.00	89.00
15	33.00	45.00	58.00	94.00	99.00
18	50.00	73.00	86.00	108.00	108.00

Source: M.S. Montgomery and W.H. Marshall, "A Simple Introduction to Solids Control," *Drilling Contractor*, (February 1982), p. 21.

The drilling fluid represents a significant portion of the total well cost (Table 5-1). One estimate is that the initial cost of a 2,000-bbl system of 15 lb/gal water-base mud is about $90,000.[10] When the well is complete, the cumulative cost of maintaining a given mud system is considerably higher. The incentive for designing better muds and using them more efficiently is obvious.

References

1. Leonard, Jeff. "What's Happening in Drilling." *World Oil*. (November 1981), p. 19.
2. Messenger, Joseph U. *Lost Circulation*. Tulsa: PennWell Publishing Co., 1981.
3. *Pocket Guide for Mud Technology*. Imco Services, a Division of Halliburton Co.
4. Lorenz, Howard. "Field Experience Pins Down Uses for Air Drilling Fluids." *Oil & Gas Journal*. (12 May 1980), p. 132.
5. Vieaux, G.J. "Heavyweight Material Reduces Well Costs." *Oil & Gas Journal*. (21 September 1981), p. 157.
6. Langenkamp, Robert D. *Handbook of Oil Industry Terms and Phrases, 3rd Edition*. Tulsa: PennWell Publishing Co., 1981.
7. Roseland, Dave. "Rules of Thumb Aid Mud Solids Control." *Oil & Gas Journal*. (31 March 1980), p. 142.
8. Liljestrand, Walter E. "Degassers Work by Bringing Bubbles to Surface." *Oil & Gas Journal*. (25 February 1980), p. 112.
9. Kelly, John, Jr. "Distinguished Author Series: Drilling Now." *Journal of Petroleum Technology*. (December 1981), p. 2293.
10. Montgomery, M.S., and W.H. Marshall. "A Simple Introduction to Solids Control." *Drilling Contractor*. (February 1982), p. 21.

6 How the Hole is Drilled

THE less information available on the area in which a well is to be drilled, the more important is well planning.

When drilling a development well in an established field, planning is much more routine. Information gained from drilling earlier wells in the field is used as the basis for casing design, drilling fluid design, bit selection, and other phases of the well plan.

After a number of wells have been drilled in a given area, well planning is greatly simplified. Casing sizes and mud programs may become almost standard for that area if the design has worked well in a number of wells. Often there are only minor changes in successive wells. Of course, the depths at which each casing size is installed will vary slightly, since formations do not lie parallel to the earth's surface. And casing depth is usually chosen so that it will be installed at a specific point in a given formation.

Even though several wells have been drilled with few problems in an area, it is still necessary to plan each well carefully and to be prepared for unexpected conditions.

Planning for an exploratory well or wildcat is much more complex, and more allowances must be made for unexpected drilling conditions. The farther the proposed well is from previously drilled wells, the less certain the drilling engineer can be of conditions that will be encountered.

In the case of a rank wildcat, very little precise information may be available for planning. The exploration team, however, because it recommends that a well be drilled there, has determined the general nature of the formations that will be encountered. Seismic data taken during the exploration phase has alerted the explorationist to the possibility of oil or gas in a particular zone. The depth of this zone is known, and the depth and some characteristics of other intervals the well will penetrate are known. Also, in many areas of the world, considerable general information is available on major geologic basins.

Well planning

Without accurate information on depths, formation pressures, and other factors that would be available when drilling in an established field, the drilling engineer must plan an exploratory well using the best information available. Where precise data are lacking, he uses guidelines that have been established for planning exploratory wells. These guidelines make allowances for the lack of exact information and may provide safety factors in several phases of the plan.

For instance, the casing design may include provisions for running an extra string of casing if a situation occurs during drilling that may call for isolating a particular formation that gives unexpected problems. If not needed, however, that string of casing will not have to be run. Or the drilling fluid plan may call for slightly heavier mud through certain intervals than would be used if conditions in those intervals were known precisely.

In selecting a rig for an exploratory well, it may be appropriate to choose one with extra capacity that might not be necessary if drilling conditions could be more accurately predicted, as in the case of a development well. Perhaps additional well control equipment or equipment to monitor the drilling operation may be included in the plan that would not be necessary when drilling where experience had shown such equipment was not needed.

Sources of information. In well planning, particularly in exploratory wells, all available information is studied. Well logs from surrounding wells or from wells with similar characteristics are an important source of information on the depths of individual zones. Logs also describe the formations penetrated in the well on which they were run, including porosity, permeability, and thickness.

Since virtually all wells drilled are—and have been for many years—logged with at least one type of tool, the records provide a valuable information library in many areas of the world. Of course, in an exploratory well far from previous drilling, conditions may be quite different from those described in such logs. But logs from previously drilled wells are often a key source of information when planning any drilling program.

When available—again, primarily when wells have been drilled previously within a reasonable distance from the one being planned—

drilling records from completed wells are a valuable source of data for planning a drilling program. On almost all wells, accurate records are kept of the type of bit used, how fast it drilled, and what its condition was when it was replaced. Drilling records also contain valuable information on the type of mud used, problems that occurred during drilling, how those problems were solved, and other data. They show at what depths casing strings were set, how much cement was used, and how long the cement was allowed to cure before drilling resumed. Data on what types of tools were run in the drill string to prevent hole deviation or other problems are also included in the record from a completed well.

All of this information is valuable in well planning: in selecting the bits to be used, where to expect casing will need to be installed, and what special problems to prepare for. These data also help estimate how long it will take to drill the well and how much the project will cost.

In most companies, approval to drill a well involves preparing a cost estimate or AFE (authorization for expenditure). This can only be done near the end of the well planning stage after the casing design has been made, a mud program has been developed, a rig has been selected, and an estimate has been made of how many days will be required to drill the well.

There are, of course, many other sources of information that can be used in well planning. Technical articles are often available that detail how a well in another area was drilled under conditions expected in the well being planned. Another good source of information in well planning is the companies that provide drilling equipment and services. For instance, many mud supply/service companies have experience in all types of drilling worldwide. They have collected an impressive amount of data on what type of drilling fluid works best under certain conditions. In addition to analyzing field data, such companies conduct continuing research aimed at developing new products to solve specific drilling problems.

Similarly, bit supply companies collect data on how their products have performed in all types of drilling. These data have been studied in detail to help determine how to apply each type of drilling bit better.

Making use of all information available is a key to good well planning. And the farther the proposed well is from existing wells, the more important planning is in drilling a trouble-free well.

Casing design. An important step in preparing a well plan, particularly for an exploratory well, is the casing design. The casing program for a well consists of several different sizes and weights of casing, set at various depths in the hole (Table 6–1).

TABLE 6–1
Casing dimensional data

Outside diameter (in.)	Nominal weight (lb/ft)	Inside diameter (in.)	Wall thickness (in.)	Inside area (sq in.)	Outside area (sq in.)	Cross-sectional area (sq in.)
4½	9.50	4.090	0.205	13.14	15.90	2.76
	10.50	4.052	0.224	12.90		3.00
	11.60	4.000	0.250	12.57		3.33
	13.50	3.920	0.290	12.07		3.83
5	11.50	4.560	0.220	16.33	19.64	3.31
	13.00	4.494	0.253	15.86		3.78
	15.00	4.408	0.296	15.26		4.38
	18.00	4.276	0.362	14.36		5.28
5½	14.00	5.012	0.244	19.73	23.76	4.03
	15.50	4.950	0.275	19.24		4.52
	17.00	4.892	0.304	18.80		4.96
	20.00	4.778	0.361	17.93		5.83
	23.00	4.670	0.415	17.13		6.63
6⅝	20.00	6.049	0.288	28.74	34.47	5.73
	24.00	5.921	0.352	27.54		6.93
	28.00	5.791	0.417	26.34		8.13
	32.00	5.675	0.475	25.29		9.18
7	17.00	6.538	0.231	33.59	38.49	4.90
	20.00	6.456	0.272	32.74		5.75
	23.00	6.366	0.317	31.83		6.66
	26.00	6.276	0.362	30.94		7.55
	29.00	6.184	0.408	30.04		8.45
	32.00	6.094	0.453	29.17		9.32
	35.00	6.004	0.498	28.31		10.18
	38.00	5.920	0.540	27.53		10.96
7⅝	24.00	7.025	0.300	38.76	45.66	6.90
	26.40	6.969	0.328	38.14		7.52
	29.70	6.875	0.375	37.12		8.54
	33.70	6.765	0.430	35.94		9.72
	39.00	6.625	0.500	34.47		11.19
8⅝	24.00	8.097	0.264	51.49	58.43	6.94
	28.00	8.017	0.304	50.48		7.95
	32.00	7.921	0.352	49.28		9.15
	36.00	7.825	0.400	48.09		10.34
	40.00	7.725	0.450	46.87		11.56
	44.00	7.625	0.500	45.66		12.77
	49.00	7.511	0.557	44.31		14.12
9⅝	32.30	9.001	0.312	63.63	72.76	9.13
	36.00	8.921	0.352	62.51		10.25
	40.00	8.835	0.395	61.31		11.45
	43.50	8.755	0.435	60.20		12.56
	47.00	8.681	0.472	59.19		13.57

TABLE 6-1 Continued

Outside diameter (in.)	Nominal weight (lb/ft)	Inside diameter (in.)	Wall thickness (in.)	Inside area (sq in.)	Outside area (sq in.)	Cross-sectional area (sq in.)
	53.50	8.535	0.545	57.21		15.55
	32.75	10.192	0.279	81.59	90.76	9.17
	40.50	10.050	0.350	79.33		11.43
10¾	45.50	9.950	0.400	77.76		13.00
	51.00	9.850	0.450	76.20		14.56
	55.50	9.760	0.495	74.82		15.94
	42.00	11.084	0.333	96.49	108.43	11.94
11¾	47.00	11.000	0.375	95.03		13.40
	54.00	10.880	0.435	92.97		15.46
	60.00	10.772	0.489	91.13		17.30
	48.00	12.559	0.330	123.88	140.50	16.62
	54.50	12.459	0.380	121.91		18.59
13⅜	61.00	12.359	0.430	119.97		20.53
	68.00	12.259	0.480	118.03		22.47
	72.00	12.191	0.514	116.73		23.77
	65.00	15.062	0.375	178.18	201.06	22.88
16	75.00	14.936	0.438	175.21		25.85
	84.00	14.822	0.495	172.50		28.56
18⅝	87.50	17.567	0.435	242.37	272.45	30.08
	94.00	18.936	0.438	281.62	314.16	32.54
20	106.50	18.812	0.500	277.95		36.21
	133.00	18.542	0.635	270.02		44.14

Source: "Baker Calculations Handbook," Baker Packers Completion Division, 1981

A string of casing is the total length of casing of a certain outside diameter that is run in a well during a single operation. A string may consist of more than one weight of casing (weight in pounds per foot) and more than one grade, even though it all has the same outside diameter.

There are three basic casing strings in most wells: the surface string, which protects fresh-water sands; the intermediate or protection string, which prevents caving and makes drilling easier; and the production string, through which the well is completed, produced, and controlled.

There are, however, more than three casing strings in many wells. There is often more than one intermediate string, for instance, when individual zones must be isolated because of high pressure or severe lost circulation.

The basic casing strings have their upper end at the surface and are cemented in place. If additional strings are required, they too will normally have their upper end at the surface.

One type of casing string does not extend to the surface. The *liner* is used to isolate or case off a section of hole below casing that has already been installed. It extends only a short distance into the bottom of the casing string above and does not extend all the way to surface.

A liner can isolate zones of high pressure or lost circulation, or it may be needed for other purposes. A tieback string is sometimes used in conjunction with a liner. This is a string of casing that extends only from the top of the liner to the surface. It is usually the same size as the liner and is used to protect upper strings of casing or to solve problems that may have occurred in those upper casing strings.

One difference between a liner/tieback combination and a conventional casing string is that the liner is run and cemented in one operation and then the tieback string is run in a separate operation. In addition, the weight of the liner is supported by a liner hanger in the bottom of the casing string immediately above the liner; then the weight of the tieback string is supported at the surface by the casing hanger. The result is that only the weight of the liner or the tieback string must be handled at any one time by the rig, not the weight of the full length of the liner/tieback combination.

This consideration can be a factor when it is necessary to install casing at great depths where the weight of a conventional casing string—one installed in a single operation to total depth—is great. Using a liner/tieback combination reduces the amount of casing weight that must be handled by the rig.

The casing design for a well to be drilled is begun by determining what is needed at the bottom of the hole and working up the hole to the surface. The size of each higher casing string is based on the smallest-size casing through which a bit can be passed to drill the hole below. When drilling the well, casing strings obviously are installed in the opposite sequence, beginning with the topmost string.

The estimated depth of the expected producing zone will determine the approximate setting depth—the depth at which the bottom of an individual string will be placed—of the lowermost string of casing, the production string. Knowing this depth, the controlling factor in selecting the proper size of each string in the rest of the casing program is the size of the production string.[1] Key considerations in choosing the size of this bottom string include the following:

1. As hole diameter increases, so will the cost of drilling. The larger the casing, the larger must be the hole in which it will be installed. Of course, the cost of drilling a larger hole must be considered in light of other factors.

HOW THE HOLE IS DRILLED 111

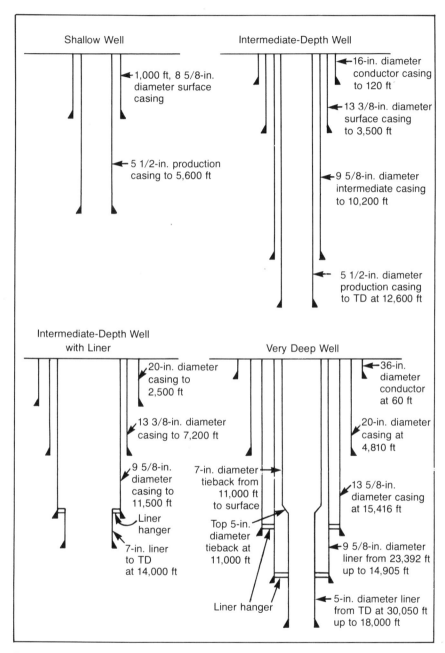

Fig. 6-1 Example casing designs.

2. If the well will require pumping rather than flowing by itself, either initially or later in its producing life, the bottom string of casing must be large enough to accommodate producing equipment. The type of fluids expected to be produced must also be considered because it will affect the type of production equipment required.
3. If high producing rates are expected, the hole must be large enough to allow these high rates without excessive pressure drop in the production tubing.
4. In some wells it is possible to produce from more than one individual zone in the same wellbore. If this is a possibility, the lower string of casing must be large enough to accommodate equipment for multizone completions.
5. The number of intermediate strings of casing also affects the size of the production string. If more than one intermediate string is required, the size of the production string will be limited.

When the size of the production string has been determined, the sizes of the other casing strings can be selected.

Along with the selection of casing sizes, the hole size (bit size) that will be drilled for each casing string must be chosen. In choosing hole size, it is necessary to consider the outside of the casing coupling as the maximum diameter that must be run in the hole. Then there must be enough clearance between this maximum casing diameter and the wall of the hole to allow for the mud cake that is built up on the wall of the hole during drilling and to allow for any equipment that will be installed on the casing to facilitate cementing operations. Allowances must also be made for possible hole curvature, which would hamper lowering the casing into the hole. It is necessary to be able to lower the casing string freely into the hole without having to force it.

Using the maximum outside diameters of the succeeding strings (working up the hole) and recommended clearances between the coupling and the hole wall, a minimum hole size can be determined for each casing string. Data have been developed on the largest bit that can be run through a given standard casing size. This is called the drift diameter of the bit. Then a bit can be chosen from standard sizes available that can drill a hole of that diameter.

This procedure only determines the bit size that will be required for each section of the hole. The type of bit must be determined using data on formation hardness, bit records from previous wells, and bit manufacturers' recommendations.

There is an added consideration in the case of exploration wells where drilling conditions and formation characteristics cannot be predicted precisely. As mentioned earlier, it may be necessary to solve an unexpected drilling problem by running a string of casing that was not anticipated in the casing plan. If an extra string must be run after the well is partially drilled and upper casing strings are installed, it means that each string of casing below that point must then be smaller in size than originally planned. Each string must be able to pass through the string immediately above it; so if a string must be run that was not expected, it reduces the size of the casing that can be run below the added string. If this possibility is not considered in the planning stages, the result might be that when the depth is reached to drill the hole for the production string, the hole that can be drilled will not be large enough to install that string. For this reason, many exploratory well casing plans include strings that are large enough to allow an additional string to be installed if necessary without reducing the size of the production string below an acceptable minimum.

The operator hopes that no major changes will have to be made in the casing plan while drilling. Actual setting depths of the strings will vary from those estimated in the plan, however, because those depths are chosen based on the expected depths of certain formations. If those formations occur higher or lower than expected, the planned setting depth of the casing string will be changed accordingly. Nevertheless, it is better to avoid having to change the size of casing or the number of strings installed after drilling begins.

From a discussion of the factors involved in casing design, it is obvious that accurate prediction of the formation pressures to be encountered while drilling is extremely important. Pressure prediction is most important in the case of exploratory wells where little drilling has been done in the area. It is also the most difficult under these circumstances.

Normal pressures in formations at varying depths are relatively easy to estimate fairly accurately. Normal formation pressures are equal to the hydrostatic pressure of the native fluids in the formation, usually fresh water or salt water. A normal-pressured formation is one that is considered an open system. If it contains fresh water with a density of 8.33 lb/gal, then the pressure increases by 0.433 lb/sq ft for each foot of depth. If such a formation existed, for example, at 10,000 ft, the pressure expected in the formation would be 4,330 psi. The hydrostatic pressure of a salt-water system increases by 0.465 psi/ft of depth, assuming a density of 9.0 lb/gal.

While normal pressures are fairly easy to estimate, the most critical factor in well planning is the presence of pressures that exceed the normal gradient. These geopressures (sometimes called abnormal pressures, although abnormal can also include pressures that are lower than normal gradients) exist because of a lack of communication between formations. Rather than being an open system, some sort of trap or seal has contained the fluids in the rock, and compaction caused by deposition of layers of sediments has caused the pressure to increase in the pore space.

There are a number of ways to determine formation pressures. Although a few can indicate the presence of overpressures prior to drilling, most can only be used to indicate abnormal pressures during drilling. These can be grouped into the following categories:

1. drilling parameters
2. drilling fluid
3. shale cuttings
4. well logging
5. direct measurement

Determining abnormal pressures is a key to maintaining control of the well at all times and to proper casing design.

Another important data element in well planning is also pertinent to designing the well casing program: fracture gradient. Fracture gradient is the pressure required to initiate a fracture in the formation. Knowledge of fracture gradients is important in planning to avoid lost circulation and to determine the setting depths of casing strings.

There are several methods for calculating fracture gradients. Generally, the calculation is theoretical or is based on equations developed using data taken during tests made while drilling.

Forces on casing. Three main forces act on casing while it is being run in the hole and after it has been installed and cemented: external pressure, internal pressure, and tension, or axial load.[1]

External pressure results from hydrostatic pressure of a fluid column on the outside of the casing as it is being run in the hole or while cementing. After installation, the pressure existing in formations may also be exerted on the casing. External pressure tends to collapse the pipe when it exceeds the pressure inside the casing.

Internal pressure results from the hydrostatic pressure of the column of drilling fluid during drilling. In addition, if it is necessary to close in the well during drilling because formation pressures have exceeded the hydrostatic pressure of the mud column, internal casing pressure may exceed that caused by the column of fluid.

Axial load on the casing is normally tension, although there are instances in which at least some of the casing can be in compression.

Tension is greatest at the top of a casing string while the casing is being lowered into the hole. Tension in the casing tends to cause the casing to fail from longitudinal deformation and it may weaken the casing's resistance to collapse (external) pressure.

Casing is classified by its outside diameter, wall thickness, grade of steel, type of joint, and length range. Weight of the casing is determined by the outside diameter and the wall thickness of the pipe. Weight can be given in pounds per foot of length or similar units. Outside diameter and wall thickness are given in inches in the English system of units. Non-English metric systems are also used. Various grades are indicated by a nomenclature that indicates the minimum yield strength of the steel. For instance, N-80 indicates the casing is made of steel with a minimum yield strength of 80,000 psi; P-110 casing has a minimum yield strength of 110,000 psi. The type of casing joints available include standard threaded and coupled joints made in accordance with specifications of the American Petroleum Institute (API) and proprietary threads manufactured by oilfield supply firms.

One of the key criteria in designing casing strings to withstand tension is the coupling where two joints of casing are connected.

Selecting the proper weights and grades of casing during well planning involves choosing the most economical weights and grades that will resist the expected forces in tension, burst, and collapse. The process is complicated by the fact that these forces vary throughout the hole. So it is not necessary to design the entire string for the maximum force in any part of the hole. Combinations of weights and grades can be selected for different intervals according to the forces in that interval. These combination strings are much more economical because the most expensive pipe is used only where forces are the greatest.

Mud program design. In the planning stage, a drilling fluid system is selected for each section of the hole. The system may be the same for a long interval, or information may be available that indicates the system will have to be changed for short intervals that present specific conditions.

In the planning stage, mud systems may be designed for sections of the hole that correspond to the sections in which different casing strings will be installed. Then only minor changes in mud properties may be needed within those intervals while drilling. Changes in the mud program after drilling begins may also be frequent and may be major depending on how actual drilling conditions differ from those expected when the well plan was prepared.

Three main considerations influence the choice of drilling fluid systems in the well planning stage; well control, drilling rate, and any known hole problems.

The well planner would like to be able to predict accurately what formation pressures will be encountered. His goal is to design a drilling fluid which has sufficient density to balance those pressures and to avoid loss of well control but which is not so heavy that it will slow penetration rate unnecessarily. If formation pressures can be balanced, fluids will remain in the formation and will not flow into the well bore. Flow of water, gas, or oil into the well bore poses the danger of loss of well control or a blowout.

However, penetration rate decreases as drilling fluid density increases. If the mud is heavier than necessary to balance formation pressures, an unneeded sacrifice will be made in drilling rate.

In planning development wells, because experience has been gained from earlier drilling in the area, it is possible to design a drilling fluid that efficiently combines these two considerations. But in exploratory wells, only very general data may be available on what formation pressures to expect. In this case, it is often necessary to design the drilling fluid to ensure formation presures will be balanced. The result may be a mud that is heavier than might be used if more accurate formation pressure information were available. But the need for protection against a possible well blowout makes some sacrifice in penetration rate acceptable. If conditions warrant during drilling, the density of the fluid can be reduced during certain intervals if it is apparent that the added safety factor is unnecessary.

It is more difficult to reduce the density of the drilling fluid than it is to increase it. Drilling fluid density is reduced by removing weighting material with the mechanical separation equipment in the surface mud system or by diluting the mud stream.

In general, it is desirable to use the lightest mud possible to as great a depth as possible before weighting up. Because the occurrence of high-pressure zones normally increases with depth, a drilling fluid plan usually will call for drilling the top part of the hole with clear water or brine. Then the lower portion of the hole may be drilled with a weighted mud system.

After high-pressure zones have been penetrated, a casing string may be installed to isolate those zones so the mud density can be reduced again. This may be done to avoid damage to the producing zone or for other reasons.

Any known hole problems are also considered in designing the mud system in the planning stage, including hole sloughing, swelling clays, thick salt sections, or the tendency of a zone to cause stuck pipe. These problems may require that the mud system contain special materials while drilling certain intervals.

Selecting the rig. The casing plan and the drilling fluid program provide essential data for selecting the rig to drill the well. Other information also goes into the rig selection process, not the least important of which is the type and number of rigs available in the area.

One way to rate drilling rigs is by their hoisting capacity, how much drill pipe or casing the derrick and its hoisting system can lift or lower into the hole. With the depth of the well known, it is possible to determine the weight of the drill string when the well is at total depth. Using data in the casing plan, the weight of the heaviest string of casing can also be determined.

Often, the longest string of casing to be installed is not the heaviest string. A shorter intermediate string may weigh more because of its larger size than a longer length of smaller-diameter casing.

The largest total weight to be handled—either drill pipe at total depth or one of the casing strings—determines the minimum requirements for rig hoisting capacity. Some additional capacity above this minimum is usually specified so the rig can exert an additional force, or overpull, on the drill pipe string if it becomes stuck.

Other rig components are evaluated or specified based on other parts of the well plan. Drilling fluid system equipment, for instance, must be capable of handling the mud required for the well. With casing sizes, hole sizes, and depths known, the volume of mud that must be circulated during each interval of hole can be determined. Also, the composition of the planned drilling fluid system, along with the type of formations and hole sizes to be drilled, indicate the amount and type of solids removal equipment needed. The proposed mud system may also dictate the use of other mud treating equipment. The volume of mud to be handled will help determine the number and size of mud tanks, or pits, required.

Special rig equipment may also be needed for an individual well. Of course, environments such as offshore drilling involve an additional list of criteria that must be satisfied by the rig chosen for the job.

A rig that provides the capacities needed and that can be equipped with any special equipment required is chosen from those owned by drilling contractors that are available or that will be available when the well is scheduled to begin.

Special situations. The basic elements of the well plan just discussed are common to all wells, whether drilled onshore or offshore and regardless of the type of rig used. Many wells, however, have special conditions that must be included in the plan. Examples of such situations are offshore drilling where plans must be extensive for resupply and other logistics and Arctic drilling where special equipment must be designed for low-temperature service. Wells that are expected to encounter hydrogen sulfide, a highly toxic gas, also require that special equipment be included on the rig and that crews be specially trained.

Moving in and rigging up

Each well that is drilled must be located accurately with respect to the appropriate survey system. For land wells the location, as the well site is called before the well is drilled, is surveyed and marked with a stake. The site for an offshore well is marked with buoys after an accurate survey has been made to pinpoint its position.

After the location is staked for a land well, it is usually necessary to prepare the site. This normally consists of leveling the area and constructing a pad of gravel or other material that will support the rig and other equipment.

In addition to the drilling pad, it is usually necessary to construct a road from the nearest existing road to the drilling site. During drilling, there will be considerable heavy traffic to and from the rig—trucks hauling pipe and casing, cementing equipment, and fuel—and a firm, all-weather road is necessary to insure that supplies can be delivered on time. The expense of building such a road could be small, compared with the cost of shutting down the rig for several days because supplies or equipment could not reach the rig.

The rig is the first equipment moved to the site. If the well is to be shallow, the rig may be truck-mounted. The truck will be positioned so the rig can be raised over the location stake. In this case, rigging up—getting the rig ready to drill—might take only a few hours. Rigs capable of drilling deeper wells will be moved to the location in a number of truck loads, and rigging up could take as long as 2 to 3 weeks, depending on the size of the rig and the equipment it includes.

Rigging up includes placing the equipment modules—mud pumps, drawworks, power generation equipment—in position and connecting piping systems and electrical wiring. The mast, which is normally moved to location in a horizontal position, is then raised to a vertical position over the point where the well will be drilled.

Moving in and rigging up on a well location on land is only one of several possible situations. It can be relatively easy where the road network is well developed; or it may be difficult, as in jungle areas where all equipment must be flown in by helicopter.

Moving to an offshore location may involve towing the drilling vessel as far as several thousand miles.

Drilling the well

Actually drilling the well consists of drilling several sections, or intervals, corresponding to the different casing strings to be installed. Each interval is drilled with a bit that will make a hole large enough for that section's string of casing. When the proper depth is reached, casing is installed. Cement is pumped down the casing and out around the bottom of the casing so it can fill the annular space.

When the cement has reached the surface in the annulus, pumping of cement is stopped. In some cases, the annulus is not filled all of the way to surface, either purposely or because hole conditions make it impossible. After the cement is in place, it is allowed to cure for a number of hours before drilling resumes below the bottom of the casing. A common term in drilling jargon is WOC, waiting on cement.

Once the cement has set, the bit size chosen for the section of hole below that casing string is run in the hole on the drill pipe. There usually are a few feet of cement in the bottom of the casing that must be drilled out before reaching the formation and beginning the hole for the next string of casing. So the size of the bit is selected so it will be small enough to pass through the casing just installed and large enough to drill a hole into which the next-lower string of casing can be installed.

Succeeding sections of the hole proceed in this general sequence: drill the hole to casing depth, run casing, cement casing in place, wait on cement, start drilling the hole for the next string of casing (Fig. 6–2). Of course, the procedure is not that simple. Problems can occur during drilling, including flows of formation fluids, hole deviation, sloughing shales, and equipment failure such as parted drill pipe or lost bit cones. Problems can also occur while running casing. In a crooked hole the casing may not lower freely. And during cementing, lost circulation can occur just as drilling fluid can be lost to a formation while drilling.

Even if the operation goes according to plan, there are many operations necessary that are not included in the general sequence of events listed earlier. For instance, the hole required for the topmost strings of casing may be quite large, and it may be impractical to drill that large a hole in one pass. If this is the case, the plan may call for drilling that portion of the hole with a bit smaller than the hole eventually has to be then reaming the hole to the necessary diameter. Although two passes

120 FUNDAMENTALS OF DRILLING

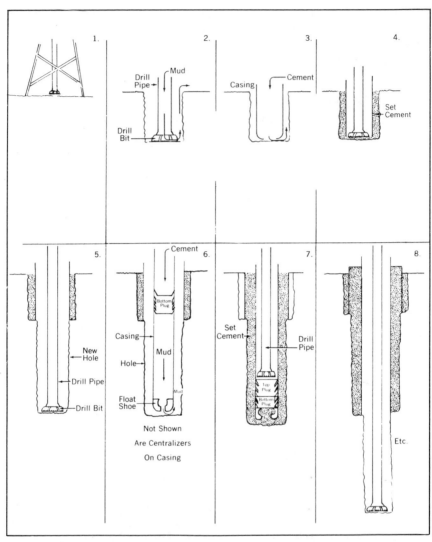

Fig. 6-2 Casing/cementing procedure. (courtesy The Western Co. of North America)

are made over the same depth interval, this approach may not take any longer than if an attempt were made to drill the hole with the larger bit. Very large and very small bits are often less durable and cause more hole problems than bits in the intermediate-size range.

Another procedure involved in drilling an interval for a string of casing is hole conditioning, discussed briefly in chapter 5. It may be necessary to circulate drilling fluid after drilling has been stopped at the casing setting depth in order to prepare the hole for the cement. The amount of conditioning required depends on whether or not the formations penetrated are stable and on other factors.

Time spent preparing the hole and the hole wall for cementing is usually considered a good investment. A poor cement job can cause problems while drilling the remainder of the well and throughout the producing life of the well.

During the drilling of each interval of hole, it may be necessary to perform a directional survey a number of times, especially in areas where hole deviation is a problem. Such surveys may be conducted as each 100-ft interval is drilled. They may be made more often if special tools or techniques are used in an attempt to correct hole deviation or to start the well along a different path. The effectiveness of these attempts is checked frequently with a directional survey. Of course, frequent surveys are made in directional wells to insure the well bore is proceeding along the prescribed path.

The survey can be conducted by lowering a tool to the bottom of the hole, usually on a wire line. The tool records the deviation of the hole from vertical. Other methods have also been developed recently that permit measurement of deviation without lowering a wire-line tool.

A record of all survey runs is kept on each well, and a three-dimensional plot of the path of the hole can be made from these data.

Running and cementing casing. After the hole is in satisfactory condition the drill string is removed from the hole. Then the crew begins the procedure for running and cementing the casing.

Each length or joint of casing is threaded on each end and is connected to the next joint by a coupling. The casing arrives at the rig with a coupling factory-installed on one end of each joint.

The casing is laid down horizontally on pipe racks adjacent to the rig. When being installed, each joint is moved to the rig floor, lifted to a vertical position, and lowered into the hole. The first (bottom) joint is fitted with a cementing shoe that lets cement flow out the bottom of the casing but not flow back in.

The first joint is lowered into the hole so its top is at the working level above the rig floor. Another joint is moved into the derrick, lifted vertically, and screwed into the top of the preceding joint with power casing tongs (Fig. 6–3), often supplied by a firm other than the drilling

122 FUNDAMENTALS OF DRILLING

Fig. 6-3 Power casing tongs are used to connects joints of casing. (courtesy Hughes Tool Company)

contractor. The casing is again lowered to a position with the top just above the rig floor and the next joint positioned. This sequence of adding a joint to the top of each preceding joint continues until the entire length of the casing string has been lowered into the hole.

During this process, equipment may be added to the outside of some joints of casing to facilitate cementing. For instance, centralizers are often used to help the casing center itself in the hole so cement will be placed completely around the pipe. Scratchers may also be installed on the casing to contact the wall of the hole to improve the chances of obtaining a good bond between the cement and the hole wall.

Prior to the time to cement casing, the company specializing in cementing oil and gas wells assembles the necessary supplies and equipment on location, including pump trucks, mixing equipment, and bulk cement (Fig. 6–4). Cement containing any desired additives is mixed with water at the rig site using automatic mixing equipment and sophisticated quality control. Then pumping equipment mounted on specially built trucks is connected to the casing.

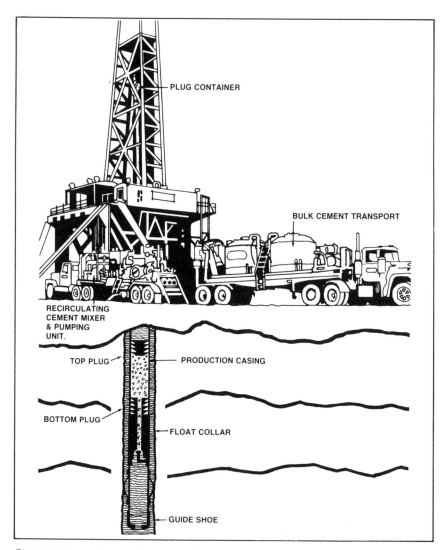

Fig. 6-4 Cementing equipment on location. (courtesy Halliburton Services)

The amount of cement has been calculated before the job begins, based on the size of the hole and the size of the casing. Logs may also be run prior to cementing to determine if the hole is in gauge—has no enlarged sections or washouts. This information is important when calculating the volume of cement to be pumped. Excess cement above the volume calculated as necessary to fill the annulus is usually pumped, the amount depending on the depth of the casing and on hole conditions. Also prior to the cementing operation, the cement is tested to determine how much time is available for pumping before the cement begins to set (pumping time) and compressive strength.

Since cement would set if left in the casing when pumping stopped, drilling fluid is pumped behind it to force all of the cement out of the casing and into the annulus. It is difficult to tell exactly when the last cement leaves the casing through the cement shoe. However, the use of cementing plugs (Fig. 6–5) helps separate the cement from the mud and gives a better indication when all of the cement is out of the casing.

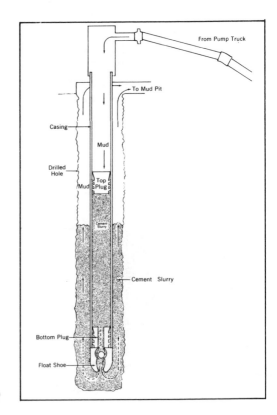

Fig. 6-5 Typical cementing operation. (courtesy The Western Co. of North America)

The advisability of running cementing plugs is not universally agreed upon, but those in favor of running them say the bottom plug that is run ahead of the cement wipes the inside of the casing free of mud that may adhere to the casing walls and contaminate the cement as it moves through the casing. When the bottom plug reaches the bottom of the casing string and stops above the cementing shoe—called "bumping the plug"—a rupture disk fails, letting cement be pumped through the plug and the cementing shoe and into the annulus.

A top plug may be run right after the cement to separate the cement from the mud that will displace the cement out of the casing. In addition to separating the mud from the cement, this plug can indicate when all cement has been displaced out of the casing. When cementing plugs reach the cementing shoe, mud pump pressure increases, indicating the plug has reached the bottom of the casing.

After the cement has been displaced around the casing, pumping is stopped. The casing may be raised or lowered slightly at this time so it is in the correct permanent position. Then cement is allowed to set or cure. After it has cured properly, the weight of the casing that had been suspended by the rig is transferred to permanent surface equipment, and some of the weight of the casing is supported by the cement.

It is common in shallower strings of casing to attempt to circulate cement, fill the annulus all of the way to the surface. In deeper strings, this is not always necessary. In some cases where circulation is desired, it may not be achieved because of hole problems.

It is often necessary to know how high the cement was displaced in the annulus if complete circulation was not obtained. The top of the cement can be determined using a temperature log that records the temperature in the hole as it is lowered down through the pipe and retrieved. Cement gives off heat as it cures, and the higher temperature recorded on the log indicates that cement was behind the casing at the depth indicated on the log.

The cemented casing string is also pressure-tested to insure integrity. Then blowout preventer equipment is installed on the casing string at the surface to control well pressures while drilling the next interval of hole.

With testing complete and blowout preventer equipment in place, drilling of the next section of the hole begins by drilling out of the casing just installed. The drill pipe string with the proper-size bit for the next section of hole is lowered to the bottom of the casing. The casing shoe at the bottom of the casing contains material that can be drilled out with the bit. When this is done, the bit begins to drill virgin formation below the casing.

This general procedure is used in running and cementing the other casing strings in the well, with variations as required by depth and hole conditions.

During the drilling of a well, a procedure called *squeeze cementing* may be required to repair damaged casing, to repair a primary cement job in which channels developed, or to solve another problem that prevented the primary cementing operation from achieving the desired results. Cement is placed by isolating the zone to be squeezed with packers or plugs then pumping cement down tubing or drill pipe.

Squeeze cementing is also used for excluding water from a producing formation or isolating a depleted producing interval so a new interval can be completed.

The techniques and the materials used for squeeze cementing vary widely, depending on the application, and many considerations must go into planning the operation. A variety of tools are available designed specifically for squeeze cementing operations.

Monitoring the drilling operation. Ideally, each interval of the hole proceeds as outlined. Many wells are drilled relatively trouble-free. In others, however, serious and time-consuming problems can occur. They may even be so serious that a portion of the hole must be abandoned and the hole side-tracked in a new direction. There have been cases in which problems were so severe that it was more practical and economical to abandon the entire hole, move the rig a short distance away, and begin a new hole. These cases are rare, but they indicate that severe problems can develop. Time spent in planning to avoid such problems is time well spent.

Problems are also avoided by watching drilling condtions closely at all times during drilling. One of the most important sources of information about conditions being encountered by the bit is the drilling fluid and the rock cuttings it brings to the surface.

Mud monitoring equipment has become very sophisticated. The amount of sophistication called for depends on the type of well. For instance, a deep exploratory well where formation properties and drilling conditions cannot be predicted accurately would call for more extensive monitoring capability.

In most wells, cuttings brought to the surface are analyzed continuously to determine the type formation being drilled. An analysis of these cuttings indicates when it is time to stop drilling and install casing, since the well plan likely calls for each casing string to be set a certain depth into a specific formation. Cuttings analysis also indicates whether or not the drilling fluid is formulated properly.

Often, the mud stream is also monitored for gas intrusion. Gas-cut mud, drilling fluid that contains entrained gas, can be a warning of a possible blowout or loss of well control. Some gas can often be tolerated in the mud stream. If the volume is small, it may be handled without danger of loss of control. For instance, degassers are often included in the surface mud treating system to remove gas entrained in the drilling fluid.

Other circulating system and mud conditions are also monitored continuously to gain information on what is happening downhole, including mud pump pressure, mud pit volume, and mud solids content.

In addition to the drilling fluid, other drilling conditions are monitored constantly during drilling. Penetration rate, for example—how fast the bit drills a given distance—is a valuable source of information. First, it tells the condition of the bit. A new bit drills faster than a worn bit; to drill the hole at the lowest cost, the bit must be replaced before penetration rate falls too low. On the other hand, it is undesirable to pull a bit out of the hole green, or while it still has significant life remaining. Pulling a bit that is only slightly worn involves the time and expense of pulling the drill string out of the hole and lowering it back to bottom. This added cost becomes greater as the depth of the hole increases.

For this reason, the drilling engineer must closely monitor bit performance using the drilling cost formula. The formula accounts for drilling time and for bit cost and trip cost. Using the formula as a guide helps replace the bit at optimum wear conditions.

A sudden drop in drilling rate, combined with other indicators, may indicate damage to the bit other than wear. Changes in penetration rate indicate changes in formation characteristics. When the bit penetrates a zone of low porosity and drills into a zone of higher porosity, penetration rate will increase. A drilling break, a significant increase in drilling rate, could indicate a formation that contains fluids. The driller must then be alert for flow of fluids from this zone into the wellbore which, if uncontrolled, could result in loss of well control.

The zone in which the drilling break occurs may not be a zone that could cause loss of control. But along with other indicators such as changes in mud volume returning from the hole, the drilling break is an early warning of possible trouble.

One problem with most of the factors being monitored is that a condition is being observed at the surface. Then conditions at the bottom of the hole are being estimated from these surface values. Decades of experience have permitted a relatively accurate assessment of what is happening downhole, based on surface observations. But the industry has long felt the need for techniques to measure drilling conditions directly at the bottom of the hole. That, after all, is the goal of most rig measurements—to determine what is happening downhole, not at the surface.

Much effort has been expended to develop such devices in recent years. A number of systems have been developed, many have been tested, and a few have become commercial tools. This equipment, generally grouped under the term measurement while drilling (MWD), is likely to be more widely applied in the future.

Making trips. One of the most time-consuming procedures required during drilling is making a trip, removing the drill string from the hole and lowering it back to bottom. The time and therefore the expense of making a trip can be significant in deep wells. Minimizing the number of trips is a goal of the well planner, the driller, and the bit manufacturer.

When a bit is worn out, the entire drill string must be pulled from the hole and a new bit must be screwed on to the bottom of the drill string. Then the drill pipe is lowered back in the hole and drilling resumes.

When pulling the drill pipe out of the hole, the drill string is pulled until three joints of pipe are above the derrick floor. The connection near the rig floor is unscrewed and the three-joint section, or stand, of pipe above is placed vertically in the derrick away from the hole with the top end in a rack in the derrick.

These stands of drill pipe are about 90 ft long. When the first stand has been racked in the derrick, the drill string is again raised until the next three joints are above the derrick floor. That stand is then unscrewed at the joint near the rig floor and placed alongside the previous stand in the derrick. This sequence is continued until all of the pipe is out of the hole and a new bit can be installed on the bottom joint of pipe.

When preparing to unscrew a stand of drill pipe, slips are placed in the rotary table to support the weight of the drill string below the connection to be unscrewed. After the stand is disconnected, the drill string is raised slightly, the slips are removed, and the drill string is raised again to remove the next stand. Initial loosening of the threaded connection and final tightening when tripping in the hole is done with drill-pipe tongs (Fig. 6–6).

When tripping back in the hole with a new bit, the sequence is reversed. A stand of three joints is taken from its place in the rack and is screwed into the connection of the stand below while the weight of the drill string below the rig floor is supported by the slips in the rotary table. After the connection has been made, the drill string is raised slightly so the slips can be removed; then the drill string is lowered until

HOW THE HOLE IS DRILLED 129

the top of the stand just connected is near the rig floor. Another stand is connected and the sequence continues.

The process of adding a joint of drill pipe as the bit drills down is similar to the tripping-in sequence. But when drilling, drill pipe is added one joint at a time.

Trips may also be necessary for reasons other than to replace the bit. For instance, other tools in the drill string may have to be replaced or a special tool may need to be added to the drill string to combat a problem.

One important consideration when making a trip is to insure that the hole is kept full of drilling fluid as the drill pipe is removed. The drill pipe occupies a volume in the hole. When that volume of steel is withdrawn from the hole, the level of drilling fluid drops. Since the

Fig. 6-6 Drill pipe tong.
(courtesy Hughes Tool Company)

hydrostatic head exerted at the bottom of the hole is a function of both drilling fluid weight and the height of the drilling fluid column, the pressure available to control formation pressures at the bottom is reduced if the drilling fluid level drops.

Not keeping the hole full during trips is one of the key causes of blowouts and loss of well control.

It is best to make a trip as rapidly as possible. In general, the longer the mud remains static in the hole without being circulated, the more hole conditions can deteriorate.

It may be that when the drill string is lowered back to bottom, it will not go all the way due to solids accumulation at the bottom of the hole. It may be necessary to circulate the drilling fluid to condition the hole for a period before the bit can be lowered to resume drilling. It may even be necessary to redrill some distance before the original depth is reached if hole sloughing has occurred.

Of course, it is best to make a trip as quickly as possible purely from a cost standpoint. A trip is unproductive time, but rig costs continue.

Fishing. It is often necessary during the drilling of a well to conduct fishing operations to retrieve a piece of drill pipe or other equipment from the well bore. If the drill string breaks or parts, the lower portion must be removed from the hole before drilling can continue. Even a cone lost from a bit may have to be removed before drilling is resumed, depending on the type of formation being drilled.

Fishing is as much art as science. A wide variety of tools exists that are run into the hole on the end of drill pipe or with a wire line to retrieve all shapes and sizes of junk. Some fishing tools may even be fabricated at the well site when a unique problem arises.

It is sometimes possible to mill or grind up junk in the hole to avoid fishing. This can often be done if the material in the hole is relatively small. It may be necessary to mill on a larger piece of junk, such as a section of pipe, to prepare it so it can be caught more easily by the fishing tool. Special hard-faced bits or mills are used for this purpose.

Fishing operations and the tools used for them are highly specialized. However, there is still a considerable amount of trial and error in fishing for what may be a broken, odd-shaped piece of metal several thousand feet down in a relatively small hole. Consequently, fishing operations can sometimes continue for long periods without success.

In extreme cases, the junk has not been retrieved and has had to be sidetracked or the well abandoned. For this reason, fishing operations should be avoided if at all possible. Specialists at fishing are very skilled and their efforts have saved many wells, but fishing is unproductive time, as far as making hole is concerned, and it increases the cost of the well.

Offshore floating drilling

After an offshore well has been started in the ocean floor, subsequent drilling operations from a floating vessel are similar to drilling from a fixed offshore platform or drilling on land: much of the well planning is the same, most casing depths are determined by the same criteria, bits are selected using the same type of data, and the mud system is designed to perform the same basic functions as in a land well.

There are some unique considerations that must be included in the drilling fluid and casing design to account for the hydrostatic pressure of the column of sea water between the vessel and the ocean floor. And because a number of operations take place on the ocean floor instead of the rig floor, there are important differences between drilling from a floating offshore vessel and any other drilling operation, especially in the early stages of the project.

Moving and anchoring. Most floating rigs are towed to location by tow boats designed for rig-towing service. These tows may be as long as several thousand miles when a rig must begin a drilling program in a new area. Or a rig may have to move only a short distance between wells when several wells are to be drilled in the same general area.

A few floating rigs are self-propelled and can move under their own power. But the portion of the drilling fleet thus equipped is small.

When the rig arrives at the well location, anchors are set in a predetermined pattern around the rig to hold it in position while drilling. The anchor pattern depends on the type of rig and expected weather conditions and is designed to provide that particular rig design with the most stability. The length of the anchor chains must be considerably greater than the water depth at the site because the anchors are set some distance from the rig. Chain storage is build into the rig, and auxiliary boats are used to place the anchors. Sometimes a combination of anchor chain and cable (chain/wire) is used in deep water to reduce weight.

Beginning the well. Starting a well from a floating offshore drilling rig is more complex than starting a well on land because well control and other equipment must be installed remote from the rig floor in a marine environment. The equipment used and the installation procedure and sequence vary.[3] In general, some sort of base must be installed on the ocean floor to support the blowout preventers; a way to guide equipment that is installed on this base must be provided; and the riser must be installed to connect the ocean floor equipment to the drilling vessel. At this point, the remainder of the well is drilled in a sequence, using techniques and equipment similar to that used for other types of drilling.

A typical well to be drilled from a floating drilling vessel might follow these steps. First, a temporary guide base is lowered to the ocean floor using the drill pipe and a tool that lets the pipe be released from the guide base when the base is in place. The temporary guide base provides an anchor for four guide lines and a foundation for a permanent guide structure. The temporary guide base is weighted to hold it in place on the ocean floor.

Next, about 100 ft of 30-in. or 36-in. casing is installed to prevent sloughing and to serve as a foundation pile to support the permanent guide base and the blowout preventer stack. The hole for the foundation pile is drilled with sea water. Since the riser is not in place, fluid returns from the hole are discharged on the sea floor. Then the permanent guide structure is attached to the top of the foundation pile and both are lowered together. Arms guide the foundation pile to the hole and shear away as the pile enters. This casing is cemented in place by pumping cement down the drill pipe string, and cement returns are discharged onto the ocean floor.

Once this permanent guide structure is in place and the foundation pile is cemented, a base has been established on the ocean floor for further operations.

Then a hole is drilled to the proper depth for the typically 20-in. conductor casing. Again, drilling fluid returns are discharged on the ocean floor since no riser has been installed. The wellhead is attached to this string of conductor casing as it is run in the hole, and the casing is cemented in place with cement returns being discharged on the ocean floor.

For the next step, the subsea blowout preventer stack is intalled on top of the wellhead and is used while drilling the rest of the well.

Finally the marine riser is connected to the riser connector on top of the blowout preventer, so drilling fluid can be circulated back to the rig floor for treatment and solids removal. The installation of the riser also provides well control in case the well begins to kick or flow because the circulating system is now closed and can be shut in if necessary.

From this point, drilling the well proceeds similarly to other types of drilling: drilling continues to the next casing setting depth, casing is run and cemented, and the next interval of hole is drilled.

When setting the guide base and drilling and cementing the foundation pile, television cameras can be used to monitor operations on the sea floor. Television monitoring makes it easier to see when cement returns have reached the sea floor, for example, and to verify that equipment is placed properly.

The marine riser. To provide a flow path for drilling fluid returning from the hole in the ocean floor and to guide the bit to the hole, the marine riser is a key element in floating drilling operations (Fig. 6-7).

Flexible joints permit lateral displacement of the drilling vessel without damaging the riser by excessive bending. One flexible joint is usually installed at the bottom of the riser; another is sometimes installed near the top of the riser in deep water.

To accommodate the vertical motion of the drilling vessel without pulling the riser apart, a slip joint, or telescoping joint, is included at the top of the riser. The rest of the riser length between the drilling vessel and the sea floor is made up of sections of riser pipe joined by riser

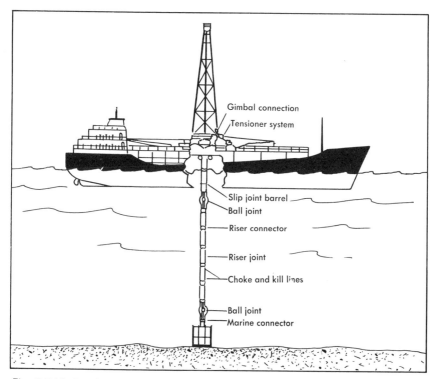

Fig. 6-7 Marine riser components. (Source: Reference 3)

connectors. Two much smaller lines, a choke line and a kill line, are usually attached to the riser—one on each side.

The diameter of the riser is determined by the size of the subsea blowout preventer to which it will be connected on the sea floor and the size of drilling tools that will have to be run through the riser while drilling.

The riser system can be subjected to severe stresses while drilling, particularly in deep water and rough weather. In addition to the constant vertical and horizontal movement of the drilling vessel, wave and current forces can be high. Some movement of the floating drilling vessel is unavoidable, but the movement is kept as low as possible to reduce the stress on the riser and the severity of fatigue effects. Logically, the steel used in the riser must be high strength and must also be able to resist fatigue failure caused by constant flexing of the riser.

When water depths increase above a few hundred feet, tension must be applied to the riser from the drilling vessel to support a portion of the riser's weight. As water depth increases, that weight increases. One expert indicates that a free-standing riser can be used in relatively calm seas in water depths to about 200 ft. In deeper water, a tensioning system should be used. A dead-weight tensioning system uses counterbalance weights on the rig that are attached by cable through a series of pulleys to the telescoping joint in the riser. In a pneumatic tensioning system, the counterweights are replaced by pneumatic cylinders that maintain the proper tension on the drilling riser as the vessel moves up and down.

Another approach to supporting the weight of the riser is the use of buoyancy material (usually made of foam) attached to the riser. Buoyancy modules can also be used in combination with a riser tensioning system to reduce the load on the tensioning system.

The riser plays a critical role in offshore floating drilling. It represents a large investment, especially in extreme water depths. Its proper design and maintenance become steadily more important as water depths increase, and damage or loss of the riser is very costly.

References

1. Craft, B.C., W.R. Holden, and E.D. Graves, Jr. *Well Design: Drilling and Production.* New York: Prentice-Hall Inc., 1962.
2. Short, J.A. *Drilling and Casing Operations.* Tulsa: PennWell Publishing Co., 1982.
3. Harris, L.M. *An Introduction to Deepwater Floating Drilling Operations.* Tulsa: PennWell Publishing Co. 1972.

7 Directional Drilling

VIRTUALLY no oil or gas well is drilled along a true vertical path. Most wells in which it is desired to drill vertically are quite close to that path, but others may deviate considerably. Hole deviation can be a serious drilling problem. It can result in the bottom of the hole being at a location other than that desired, and a crooked hole can increase the wear on drilling equipment and slow drilling progress. But in a directional well, the bit's ability to move laterally is exploited to curve the hole along a desired path.

In most wells that are planned as vertical holes, some deviation from vertical can be tolerated. As long as deviation is not severe enough to cause the well to get closer than permissible to lease boundary lines and no abrupt changes in direction exist that can damage or stick the drill string, some deviation may be acceptable. In fact, attempting to keep the hole perfectly straight would slow the drilling rate considerably and would unnecessarily raise the cost of the well.

During drilling, a constant check is kept on hole angle and position. Surveys are conducted frequently to determine the relationship of the bottom of the hole to the surface location. If hole angle or the distance the bottom of the hole is away from the surface location becomes too great, changes are made in drilling equipment or conditions to bring the hole back to the desired path.

Some types of formations have a greater tendency to cause the hole to change direction than others. Areas where these formation types exist are called "crooked-hole country". The well plan often calls for special measures to compensate for this.

During the drilling of any well, the bit may move laterally as well as vertically as it is forced into the formation by the weight and rotation of the drill string. In vertical wells the goal is to keep the bit drilling as straight as possible. Displacement from vertical may be small and still cause no drilling problems. However, the displacement may be large enough to require remedial action to prevent a hole so crooked that drilling equipment will be damaged or casing cannot be run in the hole.

There are other reasons for keeping hole deviation within acceptable limits. Producing equipment will have to operate in the well for many years and tools will have to be run in the well throughout its producing life for maintenance and repair. If a completed well is highly deviated or contains a number of abrupt changes in direction, producing equipment will wear quickly and some repair and maintenance operations may not be possible. For instance, if the well is to be produced by a pumping unit and rod pump, the rods moving up and down in the well will wear prematurely. They may also wear the casing to the point of failure.

Most offshore wells, the largest application of directional drilling, are of high enough productive capacity that equipment such as rod pumps is not needed. Special tools have been developed for maintaining directional offshore wells that can easily traverse the curved path of the well.

Directional drilling applications

Directional drilling is the process of drilling a well along a predetermined path that is not vertical. Almost any situation in which it is impossible or impractical to place the drilling rig directly above the desired location of the bottom of the well is a potential application for directional drilling. These applications include the following:

1. offshore development drilling from a fixed platform
2. onshore when drilling several wells from a central location because of environmental or other considerations
3. drilling to reach oil and gas reservoirs located under natural or man-made obstacles
4. solving certain reservoir problems by penetrating the reservoir at a high angle
5. drilling relief wells to gain control of blowouts

Offshore. One of the most common applications of directional drilling is offshore oil and gas development from fixed platforms (Fig. 7–1). Most such wells are drilled along a nonvertical path.

Exploration wells offshore are drilled vertically either from a bottom-supported mobile rig or from a floating drilling vessel. If oil or gas is discovered, the field is exploited by installing permanent fixed drilling production platforms from which development wells will be drilled. After all of the necessary wells have been drilled, the drilling rig

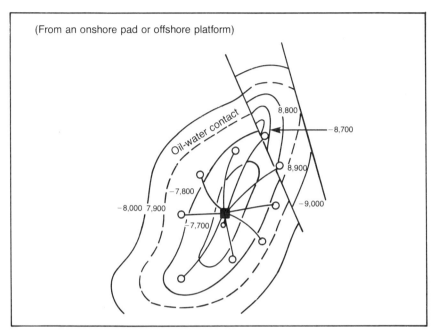

Fig. 7-1 Development by directional drilling. (courtesy Oil & Gas Journal) Source: Reference 5.

is removed and equipment is installed to treat and transport oil and gas from the field.

Fixed platforms are used for several reasons. First, since a number of wells will be needed in a relatively small area, it is easier and less expensive to drill from a fixed platform than from a floating vessel. Also, by drilling and completing the wells on the platform (wellheads and necessary valves are installed on the structure), operation and maintenance of the wells during the field's producing life are much easier. If the wells are drilled vertically with a floating drilling rig and the wellhead equipment is installed on the ocean floor, operation of the field is much more complex and costly. Finally, producing equipment may be quite complex and the technology does not yet exist to operate such equipment in a marine environment.

Even though the wells are drilled from a central platform, the bottom of the wells must be spaced out over the reservoir in a specific pattern to produce the reservoir efficiently. Directionally drilled wells can penetrate the producing zone on the desired spacing.

On the platform, a number of wellheads may be located within a few hundred square feet of space, depending on the number of wells. But the bottom of those same wells, if directionally drilled, may penetrate the producing formation over an area of several hundred acres or more.

In many offshore fields, it is necessary to install a number of drilling/production platforms because the reservoir cannot be covered properly with one. All of the points of reservoir penetration cannot be reached from a single structure. The number of wells drilled from a single platform varies widely, but slots for as many as 40 wells have been provided in a single structure.

The size, or areal extent, of the reservoir determines how many total wells will be needed in the field. When the optimum size of a single platform is determined, the number of wells per platform can be set. The depth of the reservoir is a key factor in determining how many wells can be drilled from a single platform. The formations to be penetrated, the casing program, and practical limits on the angle that a well can be deviated without excessive drilling problems also dictate how far wells can be deviated from the vertical.

As an example of how formation depth affects the number of wells that can be drilled from a single platform, assume the total overall well angle is limited to 45° from the vertical. If the formation is at a depth of 10,000 ft, then each well could reach out from the platform about 10,000 ft when it penetrated the producing zone without exceeding the 45° limit. If the formation were 16,000 ft below the surface, each well could reach a horizontal distance of about 16,000 ft from the platform without exceeding the limit on overall well angle.

Assuming both reservoirs were large enough to require several platforms, more wells could be drilled from the platform developing the 16,000-ft reservoir than could be drilled from a single platform into the 10,000-ft reservoir.

Many offshore fields produce from reservoirs at depths even shallower than 10,000 ft, and efficient exploitation can be difficult if a large number of wells is required. Several drilling/production platforms may be needed.

An overall hole deviation of 45° is only an example; it is not the maximum angle at which wells can be directionally drilled. Many wells have been drilled at much higher angles. But the greater the angle, the greater the potential for drilling problems and downhole equipment failure.

Cited as a record for horizontal displacement by directional drilling, a well completed from an offshore platform in Australia's Bass Strait reached a horizontal distance from the platform of 15,082 ft with a vertical depth of only 7,974 ft.[1]

The well trajectory was calculated using a programmable calculator and was corrected by computer input to avoid interference with existing wells. A jetting tool was used to kick off the hole; then the jet bit was replaced with a positive-displacement downhole drilling motor and a bent sub. Aluminum drill pipe was used to drill much of the hole because its lighter weight resulted in less drag on the low side of the hole and lower torque.

An earlier well in the Bass Strait had been directionally drilled to a horizontal displacement of 14,800 ft at a vertical depth of 7,950 ft.

Generally, even holes to be directionally drilled are begun vertically. The larger casing sizes installed in the top portion of the hole are more difficult to install in a curved hole. In a few cases, however, special methods have been used to begin the hole at an angle rather than vertically. Slant rigs, for example, have been used to drill directional wells by starting the hole at an angle. These rigs operate much the same as conventional rotary rigs, except that the mast is tilted at an angle for drilling. These rigs have seen only limited use and are not used to drill the bulk of directionally drilled oil and gas wells.

Another approach to starting the well at an angle is to use curved conductor pipes, the first length of casing installed. When drilling from an offshore platform installed in several hundred feet of water, it is not difficult to slant this first section of casing and allow the well to be started at an angle. In this technique, the conventional rig with a vertical mast is used.

Although neither slant rigs nor curved conductors have been used widely, they illustrate attempts that have been made to increase the horizontal displacement possible in a directional well. In normal directional drilling operations, the well is drilled vertically to a certain depth and is kicked off at an angle along the prescribed curved path. From that kickoff point, the well cannot reach as far from the surface location (the platform) as if it were started at an angle. Still, for a number of practical reasons, most directional wells are begun vertically.

Central onshore drilling sites. Offshore fields are not the only cases in which a number of wells may be drilled directionally from a central location. Central drilling pads may be used onshore to reduce the impact of drilling operations on the environment or to permit drilling and producing facilities to be enclosed so their presence will not mar a scenic area. It may also be more practical—even more economical, in some cases—to drill a number of onshore wells directionally from a central pad. For instance, in jungle areas or in the Arctic where preparation of drilling locations and roads is extremely expensive, the cost of directional drilling may be less than the cost of preparing and servicing a number of separate, widely spaced drilling sites.

Wells may also be directionally drilled where natural or man-made obstacles prevent placing the rig directly over the desired bottom-hole location. For instance, it might be necessary to penetrate a producing zone directly under a lake. Instead of building a platform in the lake to support drilling operations, it would probably be more practical and ecnomical to locate a conventional land rig near the lake and drill a directional hole.

Directional drilling from a central location has also been used to develop fields lying under residential areas. Drilling rigs and other equipment have been camouflaged to appear as modern buildings and installed on man-made islands.

Another application of directional drilling has received attention in recent years. The technique has been applied to increase recovery from certain reservoirs by drilling a hole that traverses the formation at an angle instead of penetrating it almost vertically, as is usually the case. By intersecting the producing zone at an angle, more well bore area is exposed to the reservoir for oil and gas to flow. If the hole is pictured as a cylinder passing through the reservoir, the cylinder passing through the zone at an angle has more surface area within the formation than the cylinder passing through the producing zone vertically.

There are other characteristics of oil and gas formations that make it desirable to penetrate the reservoir at an angle instead of vertically. For instance, a horizontal drain hole was drilled into a producing formation in New Mexico in an effort to maximize oil production by reducing the production of free gas in the well.[2] The hole was drilled vertically to a total depth of 6,087 ft; then the well was plugged back to 6,081 ft and 7-in. casing was set at a depth of 6,027 ft (Fig. 7–2). A special whipstock that could kick off at a much higher angle than a conventional whipstock was set just below the bottom of the casing. Using this tool and an unusual bottom-hole assembly, it was possible to turn the hole through a full right angle in only 29 ft. Then 106 ft of hole were drilled horizontally. While kicking off, directional surveys were made about every 5 ft of hole drilled.

In many directional drilling operations, it is impossible to build the angle exactly to the desired curvature then stop the angle change and proceed on a desired path. The borehole may overshoot the desired direction and have to be returned as drilling proceeds. This may happen a number of times; the directional path of the hole is not a perfect one. As a result, hole angle may be very high at a given point in the hole,

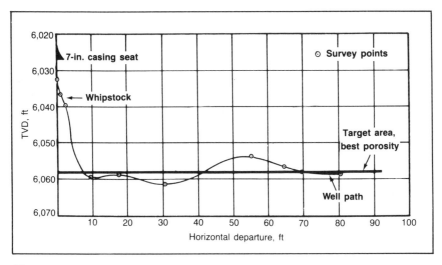

Fig. 7-2 Horizontal well path. (courtesy Oil & Gas Journal) Source: Reference 2.

compared with the overall angle from the beginning of the angle change to the end. For instance, the maximum angle recorded in the New Mexico well was 116°, meaning that the hole exceeded the horizontal path at that time and was actually aimed back at an angle toward the surface.

This example of directional drilling represents the extreme in angle building or directional change. The New Mexico well required special equipment. In addition to the bit, the bottom-hole assembly included an overgauge stabilizer and a knuckle joint. Also, a special set of articulated drill collars, or wigglies, provided the weight and flexibility needed to help the bit change direction in such a short interval. The special drill collars were cut from standard drill collar stock, and a rubber hose through the center of each section permitted mud circulation.

The operator of the field in which the drain hole technique was used planned to use the approach in drilling other wells in the field to improve well productivity. Experience gained on the first hole was expected to make subsequent drain-hole drilling easier and less costly. The cost of the first project was reported at about $500,000, roughly twice the cost of a conventional vertical hole. Changes in the procedure, however, could reduce the rig time involved, according to the operator.

In this type of operation, penetration rate in the directional portion of the hole has little effect on overall cost. Most of the rig time is spent assembling the downhole tools, making surveys to check hole angle, and performing other nondrilling operations.

The idea of drain hole drilling is not new, but it has not been widely used. Higher oil and gas prices, however, may make the approach economical in more reservoirs where the added expense could not be justified earlier. Several applications include increased intersection of fractures in tight fractured producing zones, development of better flooding patterns in enhanced recovery projects, and recovery of energy from coal beds.

Extended-reach drilling. Another concept involving directional drilling is extended-reach drilling (ERD).[3] This approach is aimed at drilling high-angle wells to obtain much greater horizontal displacement from the surface location. As the cost of producing many fields escalates, especially offshore fields that must be developed using expensive fixed platforms, such an approach to field development could reduce exploitation costs.

The dashed line in Fig. 7–3 shows what has arbitrarily been defined as extended-reach drilling. The technique would have some of the same applications as conventional directional drilling, including salt dome and fault drilling, sidetracking, drilling multiple wells from a single structure, and relief well drilling. According to Dellinger et al.,

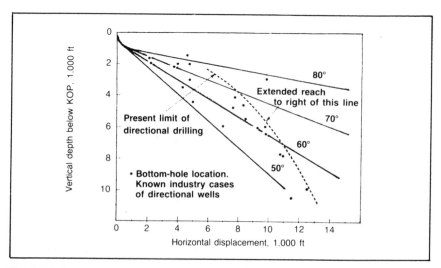

Fig. 7-3 Industry high-angle wells. (courtesy Oil & Gas Journal) Source: Reference 3.

extended-reach drilling would also have additional capabilities, including the following:

1. developing offshore reservoirs not otherwise economical
2. tapping reservoirs presently beyond technological reach
3. accelerating production by providing longer intervals in the producing formation due to the high-angle hole
4. reducing the number of platforms necessary to develop large offshore reservoirs
5. providing an alternative for some subsea completions
6. drilling under shipping fairways or to other areas presently unreachable

Much additional area could be reached from a single platform using extended-reach drilling techniques rather than conventional directional drilling methods.

The basic concept used in extended-reach drilling is similar to that used in conventional directional drilling, but the higher angles aggravate some problems common to all directional drilling. For instance, the movement of drill pipe, casing, and wire lines in the hole is more difficult. There is also an increased tendency for pipe to become differentially stuck as the hole angle increases. Other problem areas include more difficulty in keeping the hole clean and a reduced ability to control the weight on the bit and its direction.

It will be necessary to control the direction of the hole more accurately when drilling such high-angle holes. Conventional directional surveys will play an important role in this regard, but more sophisticated methods such as measurement-while-drilling (MWD) techniques will also be needed.

Relief wells. When control of a well is lost and a blowout occurs, an attempt is usually made to kill the well or stop the flow by using special techniques and equipment at the surface to regain control. Many such operations have been successful.

There are cases, however, when it is necessary to drill another well that will either intersect the out-of-control well or come close enough to it to establish a path through which fluid can be pumped into the blowing well. Relief wells drilled for this purpose pose difficult directional drilling challenges. Communication must be established between the relief well and the well it is to control, making it necessary to drill the relief well towards a precise target.

Relief wells may be drilled to the bottom of the blowing well or they may be directed at some higher point in the hole, depending on the conditions existing in the blowing well. The direction of the relief well is controlled so that it follows a path that will bring it near the wellbore of the blowing well. This path may be a single curve or an S curve in which the hole is returned to vertical as it nears the wellbore of the other well (see Fig. 7–4).[4]

Since most wells are surveyed frequently during drilling, the records can be used to calculate the location of the bottom of the blowing well using a three-dimensional coordinate system. When the surface location for the relief well is determined, the profile it must follow to intersect the other well can be calculated on the same coordinate system.

The directional plan and the characteristics of the formations to be drilled may influence the surface location of the relief well, as will the need to be a safe distance from the blowing well. So as the well bore of the relief well nears the blowing well, logging tools measure the distance and direction from one well to the other.

When the relief well has reached a point at which communications can be established with the other hole, fluids are pumped down the relief hole to control the blowing well.

Directional drilling methods and tools

The direction of the well can be changed using several methods. The angle of the hole can be increased or decreased by adjusting drilling

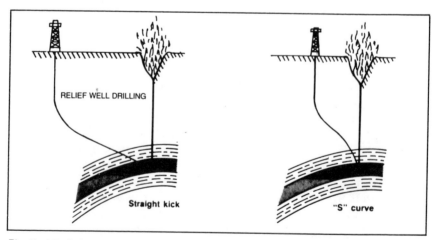

Fig. 7–4 Relief well drilling with straight kick, left, and S curve, right. (courtesy Oil & Gas Journal) Source: Reference 4.

conditions, such as weight on bit or rotary speed, or special tools designed specifically to effect a change in hole direction can be used. An understanding of the types of formations being drilled and their relationship to the earth's surface is important in choosing the tools and drilling conditions to use to drill a directional well.

Certain formations cause the bit to drill in a particular direction. If that direction is away from vertical in a vertical well, unwanted hole deviation can result. In a directional well, this tendency can make following a prescribed curved path difficult. In either case, knowing the characteristics of the formation being drilled can help compensate for the tendency of the bit to deviate from the desired direction.

Often, changes in drilling conditions can alter the direction of the hole without the use of special tools. For instance, if a hole is deviating from vertical or the curvature of a directional hole is increasing too fast, reducing the weight on the bit often reduces the rate of angle buildup. Called *dropping angle*, less weight on the bit causes it to move toward the vertical due to a pendulum effect. In addition to changes in bit weight, adjusting the rotary speed may also help achieve the desired angle change.

An important consideration in analyzing the formation for directional drilling purposes is its slope. If the bit is drilling vertically and encounters a sloping layer of rock, it may tend to drill along the plane of that rock layer rather than in the desired direction. Also, because the bit is rotating as it drills, it often tends to "walk" continually to one side, following a spiral path. If minor, such deviation from the desired path can be tolerated. But if hole angle and direction are not monitored closely with corrections made when needed, serious problems can result.

Because the hole does not change direction in only one plane, any point along the well bore must be described by three coordinates: a distance east or west of the vertical projection of the surface location, a distance north or south of that projection, and a depth from the surface. Information gained from directional surveys conducted during drilling is used to plot the path of the well bore in these three dimensions.

Adjusting bit weight or other drilling variables is a common approach to reducing the angle of the well path toward vertical. To increase the curvature of the hole or to initiate a direction change from vertical along a curved path, mechanical means are usually necessary. Of the mechanical tools specifically designed to change the direction of a hole, the whipstock is the oldest (Fig. 7–5).

The whipstock has a chisel-shaped lower end and is made so the bit is forced to one side of the hole when the bottom of the tool is embedded in the bottom of the hole. When the tool is in place, increased weight on the drill string shears a pin that attaches the tool to the drill string. The bit then drills at an angle to the previous hole. When a pilot hole has been drilled for a short distance, the whipstock is removed and drilling along the new path can resume.

Whipstocks are often used when it is necessary to sidetrack a hole out through casing or in an open hole when other methods for changing hole direction are unsuccessful.[5] The biggest advantage of this tool is that it provides controlled conditions and positive action for changing hole direction. It distributes the side force on the bit caused by the whipstock's curvature over the length of the tool. One disadvantage of the whipstock method of changing hole direction is that, after the hole has been kicked off with the pilot hole, it is necessary to make a trip to replace the smaller bit with one that will drill the desired hole size.

Positioning the whipstock is a critical opeation. It is also important that the proper weight on bit, rotary speed, and hydraulic conditions be used while drilling the pilot hole. Proper drilling conditions are equally important after the small bit has been replaced with a larger bit, especially in softer formations. For example, if mud circulating rate is in-

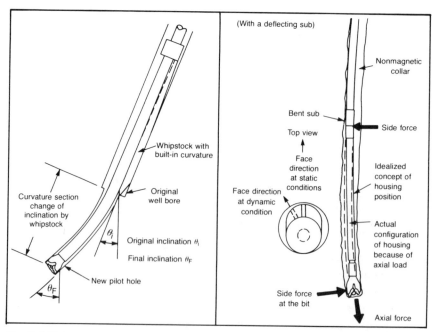

Fig. 7-5 Hole deviation by whipstock, left, and by downhole motor, right. (courtesy Oil & Gas Journal) Source: Reference 5.

DIRECTIONAL DRILLING 147

correct or improper weight on bit is used, the larger bit will not follow the pilot hole and the direction of the hole will not be changed as desired.

Much of today's directional drilling is done with downhole drilling motors (Fig. 7–6) used in combination with a bent sub. The bent sub is a section of drill pipe manufactured with a slight angle that is installed in the drill string above the bit. The built-in angle of the sub exerts a side force on the bit and causes it to be deflected from the previous direction of the hole. Bent subs provide deflections from near zero to 2½° or more.

Experts recommend making a slow change in hole direction when using a downhole motor and bent sub. It is better to use a bent sub with a smaller deflection angle and make the correction over a longer hole interval than to attempt to make a dramatic direction change in a few

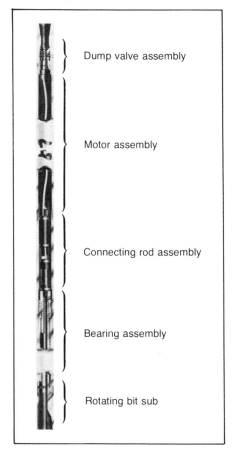

Fig. 7–6 Downhole drilling motor. (courtesy Dyna-Drill Division of Smith International Inc.)

feet. The gentler curve makes it easier to control the new direction of the hole, and fewer problems in subsequent drilling will result.

Another directional tool that operates on the same principle is a downhole drilling motor with a bent housing. The drilling motor is manufactured with a built-in angle to cause bit deflection. The angle of the drilling motor itself exerts the side force on the bit, and a bent sub does not need to be run above the drilling motor.

Of the two general types of downhole drilling motors, only the positive-displacement mud motor is available with a bent housing. Since the turbine-type downhole motor has a shaft through the tool, it is impractical to build with an angled housing. It must be used with a bent sub.

Another way to change the direction of the hole is by jetting. Drilling fluid is circulated through a bit with a large jet that is oriented in the direction desired. When mud is pumped through the large jet, the hole is eroded in the direction it is aimed. The degree of deviation that is possible with this method depends on the design of the drill string just above the bit. Direction change using this method is not as easy to control as is the case with other techniques; the formation may not erode in precisely the direction desired. When the bit is lowered while jetting, it may not follow the direction in which it was originally oriented.

One advantage of this method is that it is relatively easy to make several attempts to establish the proper hole direction by reorienting the bit and jetting again. In addition, when the desired hole direction has been established, drilling can proceed using the same bit.

The jetting bit may be made by installing a large jet in a conventional three-cone bit along with two smaller jets. There are also special jetting bits available, including a two-cone bit in which an extended jet takes the place of the third cone, and a conventional three-cone bit with a fourth large jet.

When planning a jetting operation, a key consideration is the capacity of the rig's mud pumps. This, along with other factors, determines the size of the jet that will give the best performance.

According to Millheim, the following should be considered in choosing which method to use to change the direction of a hole:

1. depth at which the change in direction will be made
2. hardness of the formation
3. experience of the directional drilling operator
4. rig hydraulics
5. rotary table condition

DIRECTIONAL DRILLING 149

6. hole fill-up problems
7. trend of geology relative to the direction of the well
8. inclination of hole at depth of correction or kickoff
9. accuracy of rig instrumentation (weight on bit, rotary speed, torque)
10. experience in the area in kicking off or changing hole direction
11. possibility of getting stuck, losing circulation, or taking a kick

Each method offers advantages in certain applications. For example, jetting is normally considered for depths shallower than 5,000 ft; other methods are better for greater depths. Adequate hydraulic conditions, such as enough pressure at the bit, is a factor in defining the practical depth limit for jetting. Use of a whipstock generally requires more experience than is needed for other methods, while the mud motor/bent sub combination requires less experience. Formation hardness is a key criterion in choosing which directional drilling method to use. For harder formations, whipstock and mud motor/bent sub techniques are most applicable. Jetting is usually the best choice in very soft and medium-soft formations.

Bottom-hole assemblies. A whipstock, downhole motor/bent sub combination, and jetting change the direction of a well. But most of the influence on hole direction is exerted by the bottom few hundred feet of the drill string, the bottom-hole assembly (BHA).

The BHA is generally defined as those components that are part of the active section of the drill string, the portion that influences hole direction, inclination, and penetration.[6] This may include the bit, drill collars, stabilizers, and special subs (short lengths of drill pipe designed for a specific purpose, such as the bent sub).

An almost infinite combination of these drill-string components is possible, and the analysis of the effect of a given bottom-hole assembly or the prediction of how it will perform is a complex procedure. Each component has different physical properties. How they will act together in a given sequence in the drill string is what the drilling engineer must determine. To complicate the matter, there is still much that is unknown about what causes bits to follow a given path as drilling progresses. Also, virtually every hole is unique. The result is that much of the data used in designing bottom-hole assemblies is empirical, obtained from drilling many wells.

In addition to the physical properties and dimensions of the various components in a bottom-hole assembly, density of the drilling fluid, inclination of the hole, hole size and curvature, position of each component in the drill string, and weight on bit influence the performance of the assembly.

Computers have been used to analyze the variables involved in the design of a bottom-hole assembly. Programs have also been written for the widely used hand-held calculators that allow the drilling engineer to design the proper drill-string assembly in the field.

The importance of the bottom-hole assembly cannot be overemphasized in drilling directional as well as vertical wells. Changes can be made in the bottom-hole assembly to help correct hole deviation without the use of one of the corrective tools described earlier. For instance, a packed-hole assembly may be used. It consists of drill-string components that fill the hole so the drill string is unable to flex or bend.

A packed-hole assembly will not always keep the hole from deviating, especially if the formations are of the type that tend to cause hole deviation. But this type of drill-string assembly may retard changes in the hole angle.

If hole angle is increasing, a pendulum assembly is often used. In this approach, stabilizers are used at a selected point in the bottom-hole assembly to cause the bit to rest on the low side of the deviated hole. This may bring the hole back toward a vertical path. To be most effective, the drilling assembly is usually operated with less weight on the bit than would otherwise be applied. Adjustments may also be made in rotary speed and in drilling fluid hydraulics to make the assembly perform as desired.

Some problems

Problems discussed earlier in relation to extended reach drilling, although more serious as the hole angle increases to the levels used in that technique, can also be problems in conventional directional drilling at more modest angles. In holes intended to be straight and vertical but which become deviated, these same problems can exist in varying degrees.

Problems that can occur in wells which deviate from the vertical—whether intentionally or not—include additional wear of the drill pipe, casing, and other downhole tools. Casing can be more difficult to lower into the hole if the hole is deviated, and the difficulty increases as the angle of the hole increases. The danger of the drill pipe sticking against the wall of the hole due to the pressure difference between the drilling fluid column and the formation can also increase in nonvertical holes.

Additional drilling problems can result if a dogleg develops in a hole. A dogleg is a relatively abrupt change in hole direction, usually measured in degrees per 100 ft, and can cause excessive stress in drill pipe, tool joints, and other drill-string components. Doglegs also increase the likelihood that the drill string will become stuck.

After the well is completed, as discussed earlier, producing operations can be hampered and operating costs can be increased as a result of hole deviation.

References
1. Chambers, Mike, and Jim Hanson. "Australian Well Breaks Horizontal Displacement Record." *World Oil*. (March 1982), p. 123.
2. Moore, W.D. III. "ARCO Drills Horizontal Drain Hole for Better Reservoir Placement." *Oil & Gas Journal*. (15 September 1980), p. 139.
3. Dellinger, T.B., W. Gravley, and G.C. Tolle. "Directional Technology will Extend Drilling Reach." *Oil & Gas Journal*. (15 September 1980), p. 153.
4. Adams, Neal. "Blowout Control–2: How to Drill a Relief Well." *Oil & Gas Journal*. (29 September 1980), p. 93.
5. Millheim, Keith. "Directional Drilling–2: Proper Application of Directional Tools Key to Success." *Oil & Gas Journal*. (20 November 1978), p. 156.
6. Millheim, Keith. "Directional Drilling–3: Here are Basics of Bottom-Hole Assembly Mechanics." *Oil & Gas Journal*. (4 Decemer 1978), p. 98.

8 Well Control and Safety

PROBABLY the most important concern during drilling is that the pressure which exists in any formation penetrated by the bit must be controlled at all times. Detailed well planning is the first step in preventing trouble while drilling. After drilling begins, constantly monitoring drilling variables, using appropriate equipment, and employing well-trained drilling crews can drastically reduce the chance of losing control of the well.

Consequences of blowouts

Failure to control well pressures can result at the very least in problems that impede drilling progress. At its most serious, failure to control formation pressures while drilling can cause loss of life, destruction of equipment, and abandonment of the well (Fig. 8–1). A well blowout can also damage the surrounding environment.

All of these consequences have caused great emphasis to be placed on the design and use of blowout control equipment, personnel training in well control, and regulations aimed at minimizing the chance of well blowouts.

Because offshore blowouts are especially troublesome and a few incidents have been particularly dramatic, they have had considerable influence on industry and government efforts to prevent both offshore and onshore blowouts. Several of these occurred in a short time span in the late 1960s and early 1970s offshore California and in the Gulf of Mexico (Fig. 8–2). These incidents were widely publicized and brought demands that offshore drilling be halted. In 1979 another blowout in the Gulf of Mexico in Mexican waters was also the subject of controversy. Fears of extensive, long-term environmental damage resulting from these and other blowouts that produced oil spills appear to have been at least somewhat exaggerated. Continuing studies of the area involved indicate that the oceans have a great capacity to recover from the effects of such accidents as long as they occur infrequently.

All of these situations are serious, and all can have serious consequences. But there are differences among types of blowouts. Uncontrolled flow from a gas zone, for instance, is often ignited and the gas burns as it flows from the well, leaving little residue on the ground or on the water around the well. Regaining control of a blowout on land is often easier because equipment can be brought to the well and operations can be conducted more readily than when the blowout is offshore.

In the early days of drilling, uncontrolled flow from a well was not uncommon; the gusher signaled success. That has not been the case for decades now, and in the last 20 years more and more industry training and technology development efforts have focused on blowout prevention.

The seriousness of loss of well control varies widely. Some wells have blown out and not been brought under control for months or more. Other blowouts are brought under control in a few days. When well control is lost, damage may be slight. But in the most extreme cases, many lives are lost and the rig and related equipment totally destroyed by fire.

Fig. 8-1 Blowout and fire destroys rig. (courtesy Oil & Gas Journal*)*

Fig. 8-2 Equipment mobilized at offshore blowout. (courtesy Oil & Gas Journal)

Pressures have been encountered while drilling that were great enough to force the heavy drill string and other equipment from the hole and up through the top of the derrick. If pressure in a formation penetrated by the bit significantly exceeds the hydrostatic pressure of the mud column and the formation is highly permeable to the flow of fluids, loss of well control can happen quickly—sometimes even before the rig crew can leave the rig floor. If the flow ignites from an engine spark or other source on the rig floor, the resulting fire can be devastating.

Blowouts are costly for those involved in the well and for the entire industry. In addition to the direct expense involved in capping a well blowout, liability for damage to the surrounding area can cost the operator and the drilling contractor additional millions of dollars. Failure to keep such drilling accidents to a minimum can bring impractical and unnecessary restrictions on exploration in certain areas and delays in developing needed energy. For these and other reasons, much attention is given to blowout prevention.

The simple answer to these problems is to develop a drilling plan, including a drilling fluid design, that will control any flow from any formation encountered. The trouble with this simple solution is that each well is different, and no standardized well design or procedure can be applied to all drilling. A drilling program must be designed that includes some contingency or safety margin to handle unexpected pressure or formation fluid flow. At the same time, the program must account for zones that may not be pressured, which call for a mud with different properties. And, the drilling fluid must let drilling proceed as fast as practical, considering other criteria.

The result is that a drilling fluid design represents the best possible compromise among the several demands on it. Sometimes, unexpected drilling conditions exceed the capability of the mud system.

When blowouts are most likely

The industry's safety record is best for frontier area drilling.[1] Though often less is known about exploratory well conditions than about conditions where previous drilling has provided experience, more precautions are apparently taken in exploratory wells and deep-water wells.

It has long been acknowledged that one of the operations during which the danger of losing control of the well is greatest is tripping the drill string out of the hole. During trips, control of formation pressure can be lost if the hole is not kept full of drilling fluid as the pipe is withdrawn. If the pipe is pulled too fast, swabbing can result. Swabbing decreases the pressure exerted by the fluid column below the bit, and formation fluids may flow into the well bore.

A 1971 study of 32 incidents in which well control was lost revealed that in 10 out of 19 blowouts that occurred while drilling development wells, the drill string was being pulled from the hole.[2] The same study showed that in only 2 out of 10 blowouts while drilling exploratory wells was pipe being tripped out of the hole. These data indicate further that the added precautions often taken in drilling exploratory wells, because conditions cannot be predicted accurately, pay off.

Well planning is extremely important in the case of exploratory wells because little or no experience in the area is available. It is also necessary to plan development wells. But after a number of wells have been successfully drilled in an area, less safety margin is required for unexpected conditions.

The proficiency of rig crews and supervisors and their ability to put contingency plans to work if drilling conditions indicate a problem is imminent are critical to avoiding blowouts. No amount of contingency planning is adequate if rig personnel cannot implement the plan quickly and properly.

In wells discussed in that 1971 study, loss of well control occurred equally as often while actually drilling. Loss of control was also reported during circulating, when drilling after freeing stuck pipe, and while going in the hole with the drill string.

Control of a well can be lost while drilling at almost any depth. The presence of gas sands at shallow depths that contain pressure higher than the normal pressure gradient for that depth has been the cause of a number of well blowouts. Overpressured zones can be found at almost any depth. Even if pressure increases according to expected pressure gradients, it is necessary to maintain an adequate mud weight to balance formation pressure.

There is, therefore, little correlation between the danger of loss of well control and depth. But the 32-well study indicated some relationship between the ratio of open (uncased) hole to total depth and the danger of a blowout. For instance, data indicated that two intervals are potentially dangerous: soon after setting casing and as the hole nears the next casing depth. Midway between two casing points, the danger of losing control is less.

Casing is often set in connection with abnormal-pressure zones to isolate such a zone from weaker ones above or below. So the troublesome formations, those likely to contain high pressure, are often located near casing setting depths.

This particular study, though by no means comprehensive, also showed that of the 32 wells on which control was lost, 20 were either plugged and abandoned, sidetracked, or plugged back to a shallower depth. Of these 20, 10 were development wells. In the 20 wells in which drilling could not continue in the original hole, about 143,000 ft of hole already drilled had to be abandoned. An estimated 68,000 ft of casing was also lost, along with approximately the same amount of drill pipe and drill collars. On three of the wells, the rig was reported destroyed and it was assumed lost on a fourth well. It was impossible to estimate the cost of fishing jobs, extra mud, and cementing jobs but the extra expense caused by loss of control was significant.

Offshore drilling can present special well control problems, and the consequences of loss of control can be more serious. In the case of development drilling from fixed platforms, for example, several wells may have been completed on the platform while drilling proceeded on others. Although production from a platform is normally not permitted while drilling is being conducted, the presence of completed wells near the drilling operation calls for special care.

Drilling from a floating mobile offshore rig can also present special hazards. For instance, if large volumes of gas escape from the reservoir

Flow of formation fluids

Fluids that can enter the wellbore and cause a well kick include oil, water, and gas. The most common occurrences involve salt water and gas. Gas, because it often enters the hole in large volumes and because it is compressible, can be one of the most difficult flows to handle. The consequences of loss of control of a gas kick can be the most dramatic.

The volume of gas depends on the pressure and temperature conditions under which it exists. As it enters the well bore from the formation in a deep hole, it is under considerable pressure. If the gas is allowed to move to the surface through the drilling mud, it will expand in volume as it moves up the hole because the pressure exerted by the mud column decreases as the depth decreases. This expansion accelerates as the gas bubble nears the surface. When the gas reaches the surface, its volume can be several times the original volume at the bottom of the hole. As the gas moves to the surface and expands, it displaces drilling fluid from the hole, causing a rapid gain in the level in the mud pits. The hydrostatic pressure of the mud column is reduced, and additional gas can enter the well bore.

Flows of salt water and oil into the well bore can be difficult to control, but a flow of gas is one of the most critical well-control situations.

Indications of flow. Because of the expansion of a gas bubble as it moves up the hole, an early indication of a gas kick is an increase in the level of fluid in the mud pits. A close watch on mud pit level by the crew, augmented by modern monitoring and alarm devices, is the first line of defense against losing control of the well during a gas kick.

Even before a significant increase in pit level is apparent, increased flow of mud from the well will signal a possible influx of fluid at the bottom of the hole. If there is doubt about whether flow has increased, stopping the mud pump will help determine the answer. If the well continues to flow after the pump is shut off, formation fluids are probably entering the well bore.

Often, the influx of fluids into the well bore—even sizable volumes of gas—does not mean a blowout is certain. It is possible to take a kick, follow the proper procedure for circulating it out using equipment designed for that purpose, and resume drilling without serious consequences.

Role of the drilling fluid. As discussed in chapter 5, the drilling fluid is a key element in controlling formation pressures while drilling. A kick occurs when the pressure in a formation penetrated by the bit exceeds the hydrostatic pressure exerted on the formation by the drilling fluid column. So one of the most important tools for well control is a crew trained in monitoring the signals that indicate reservoir fluids are entering the wellbore.

By itself, any single indicator may not tell the crew that a kick is certain. But analyzing several drilling variables can warn of a kick, and preventive action can be taken. For instance, an increase in the rate at which drilling fluid is returning from the well when pumping rate has not been increased means reservoir fluids are entering the hole.

An increase in the level in the mud pits (a volume increase) also signals that formation fluids are displacing drilling mud from the hole, providing no changes have been made in pump rates or other drilling conditions. Monitoring the volume of mud in the pits is so important that a variety of equipment is available to detect changes in mud pit volume automatically and sound an alarm if they exceed a normal range.

And, as mentioned before, if flow continues after the mud pumps have been stopped, flow of formation fluids into the well bore is indicated.

Failure to keep the hole full of mud during a trip can lead to a kick or loss of well control. The hydrostatic pressure exerted at the bottom of the hole by the column of mud is reduced as the level in the hole is lowered. Related to this is the fact that if the mud level does not fall an amount equal to the volume of steel removed from the hole when the drill pipe is withdrawn, a kick may be indicated. Formation fluids may be entering the well bore and occupying the volume previously occupied by the drill pipe, preventing the mud level in the hole from dropping as expected.

Gas entrained in the drilling fluid, gas-cut mud, may also indicate that gas from a formation penetrated by the bit is entering the hole. The volume of gas may not be large enough to cause an actual kick, so drilling often continues with some gas being brought to the surface in the mud stream and removed. In this case, it may not be necessary to increase mud weight or to initiate any well control procedures; drilling can continue with care.

Other drilling variables can warn of impending kicks. A drilling break, an abrupt change in penetration rate, can indicate that the bit has drilled into a formation that may be more likely to contain fluids under pressure. A gradual change in penetration rate is common when drilling into a slightly different rock, but many formations that contain significant volumes of fluids have characteristics that make them drill considerably faster than other formations.

The difference between a drilling break and a normal increase in penetration rate is not always apparent. Experience and a knowledge of the formations in the area are important in recognizing a drilling break.

The flow of fluids into the well bore is not necessarily a cause for alarm, and no significant changes in the drilling operation may be needed. The seriousness of such flow is closely related to the volume of fluid that is entering the hole, in addition to the pressure in the formation. The volume of fluid that will flow from the formation into the hole depends largely on the permeability of the formation rock.

For instance, if a formation is encountered that contains a pressure higher than the hydrostatic pressure of the mud column but has low permeability, the volume of fluid entering the well bore may be small. Small volumes can often be tolerated in the drilling mud, circulated to the surface, and removed on a continual basis. In such situations, drilling often proceeds without significant mud weight increases or other changes. The combination of a lighter mud weight and the pressure differential between the formation and the hole causes rock chips to be more easily broken from the rock by the bit, and faster penetration results. So maintaining the mud weight as low as possible and letting small volumes of fluid enter the hole can result in faster drilling rates than if the mud were weighted up to the density needed to stop all fluid flow from the formation.

Hydrogen sulfide. One of the most dangerous fluids encountered during drilling is gas containing hydrogen sulfide (H_2S). In addition to the dangers of any gas flow from the well, hydrogen sulfide is highly toxic in very small amounts. A relatively short exposure to it can cause death.

Failure of drill pipe and other metal drilling equipment is also rapid in the presence of H_2S. Consequently, when drilling is planned for an area in which hydrogen sulfide exists or is suspected in formation fluids, the well plan must include detailed procedures for protecting drilling equipment, the drilling crew, and the public in surrounding areas.

Hydrogen sulfide has a detectable odor at very low concentrations. But at higher concentrations it deadens the sense of smell and can no longer be detected. It is also heavier than air and will collect in low areas around the rig.

When a well is to be drilled that may encounter hydrogen sulfide, the well plan must provide for monitoring devices, both stationary and portable; protective breathing apparatus for crewmembers; and emergency equipment to treat the effects of exposure. Detailed instructions must be prepared for each crewmember, outlining his actions in case an emergency arises. The plan must also contain measures to protect persons living or working in areas surrounding the drilling operation.

Because of its detrimental effect on steel, the possibility of encountering hydrogen sulfide must be considered in planning a mud program and designing tubulars that will be used in the well. Drilling fluids are designed to inhibit H_2S corrosion as much as possible, and drill pipe and casing strength characteristics and temperature limits are selected to minimize their susceptibility to H_2S attack. The blowout preventers, a critical link in well control, must also be selected especially for H_2S service.

How kicks are handled

The occurrence of a kick does not mean that a blowout is inevitable. Indeed, kicks are handled routinely by experienced crews using the proper equipment and procedures. Relatively few kicks, in fact, result in total loss of well control. But if a kick is not handled properly, it can be serious.

A number of techniques have been used when a kick occurs. The procedures differ in detail, depending on whether the rig is floating or fixed. Arguments can be made in favor of one method over the other; but in general, procedures to follow when a kick is indicated and before a kill procedure can be initiated include these steps:

1. As soon as one of the warning signs of a kick is observed, the well should be shut in. This is done by stopping the mud pumps and closing the blowout preventer around the drill string. Before closing the preventer, it is normal to raise the kelly.
2. Most instructions call for notifying key company personnel at this stage that a kick is suspected and that the action in step one has been taken.
3. After the system has reached equilibrium—maybe a few minutes—the pressure on the drill pipe, the pressure on the casing, and the increase in the volume of mud in the pits are recorded. At

this point, calculations can be made to determine the type of fluids entering the well bore, the weight of mud needed to stop the flow, and other data necessary for implementing the kill procedure.

Increasing the density of the drilling fluid to the kill weight is necessary in order to stop the flow of fluids into the well bore. The kill weight is the mud weight that will provide a hydrostatic pressure at the formation exactly balancing the pressure in the formation.

When mud weight is given in pounds per gallon, the hydrostatic pressure at any depth is obtained using the depth and a conversion factor to convert pounds per gallon to pounds per square inch. Since the mud weight must be mixed to achieve a certain density in pounds per gallon, common usage often refers to the formation pressure in these terms. For example, a 12-lb/gal kick means the mud weight must be increased to 12 lb/gal to balance the pressure in the formation.

After the well has been shut in and the weight of mud that will be needed to kill the kick has been determined, the remainder of most procedures involves various ways to circulate out the kick.

Most of the recommended procedures for kick killing include maintaining a constant bottom-hole pressure during the procedure to prevent further flow of fluids into the well bore. Then by proper circulation sequence, proper mud weight, and the use of pressure control equipment, the formation fluids are removed from the well bore.

Variations in this basic procedure involve when the kill mud is circulated. In one approach the heavy mud is used to pump out the fluid that has entered the well bore. In another, the kick fluid is pumped out of the hole before the mud weight is increased. Still a third method increases mud weight at the same time the kick fluid is being circulated from the hole.

The wide variety of situations under which kicks occur calls for a number of modifications of this basic kick monitoring and killing procedure. For instance, in floating offshore drilling the movement of the vessel can make it difficult to determine changes in mud pit level. Constant movement of the pipe as the vessel moves in the water can present problems when closing the blowout preventers on the pipe. Since the BOP is on the ocean floor, rather than on the rig floor, its operation is more complex.

Regardless of the type of drilling, drilling conditions, or the method used, the goal of a kick-handling procedure is to circulate the fluid that has entered the well bore out of the hole and to balance the formation pressure with drilling fluid so no more formation fluids will enter.

Blowout prevention equipment

Mechanical equipment directly involved in blowout prevention includes several types of surface blowout preventers, downhole blowout preventers, and a variety of valves that can be installed in the kelly and at other points in the drill string.

Blowout preventers are used when the drilling fluid column has failed to contain fluids in the formations being drilled and it is necessary to shut the well in to prevent an uncontrolled flow.

The main types of surface blowout preventers are the annular and the ram type. Annular preventers (Fig. 8-3) consist of a steel body containing packing elements and an actuating mechanism. When it is necessary to close the preventer, hydraulic pressure from a BOP control system operates a piston that pushes the packing elements out into the hole around the drill pipe. Reversing the action of the piston opens the preventer and moves the packing elements back into the body out of the

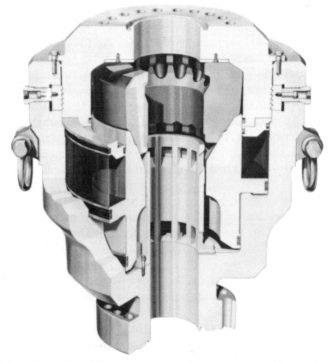

Fig. 8-3 Annular surface blowout preventer. (courtesy Hydril Mechanical Products Division)

way of pipe movement. The bottom of the preventer body is designed to be connected with other blowout preventers.

Ram-type preventers (Fig. 8–4) also have a steel body and are similar in operation. Hydraulic action forces the preventer to close and open, but the elements that close around the drill pipe are two rams. Each closes against the drill pipe from opposite sides within the blowout preventer body.

Three common types of ram-preventer rams or elements are the pipe ram, the blind ram, and the shear ram. The pipe ram closes around the drill string to close off the annulus of the well between drill pipe and casing. Its purpose is much the same as the annular preventer. The two elements in the pipe ram each have a semicircular opening to fit around the size of pipe being used. When closed, the two elements surround the drill pipe and seal off the well.

Blind rams have no semicircular opening. The end, or face, of the ram elements is flat, and the two elements close tightly together to shut in the well when no drill pipe is in the preventer.

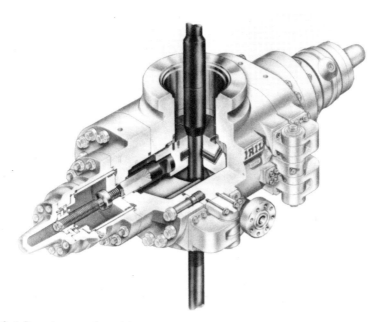

Fig. 8-4 Ram-type surface blowout preventer. (courtesy Hydril Mechanical Products Division)

164 FUNDAMENTALS OF DRILLING

The shear ram is a type of blind ram with a cutting edge on the face of each element that can shear off the drill pipe if necessary and shut in the hole above the sheared pipe. If shear rams are closed and the drill pipe is sheared off, the portion of the drill string below the rams will drop. If desired, this can be prevented by having a pipe ram preventer below the blind-ram preventer that holds the drill string after it has been cut by the shear rams. The lower pipe ram is closed first, normally immediately below a tool joint; when the pipe is cut, the drill string will be supported.

There are other variations in design of blowout preventers, but these key types are used on most wells.

More than one BOP is usually used. They are combined, one above the other, in a blowout preventer stack (Fig. 8–5). The BOP stack is located under the rig floor, except in the case of an offshore floating rig where the blowout preventer stack is located on the ocean floor. The number

Fig. 8-5 Blowout preventer stack includes annular unit, top, and ram-type BOPs below. (courtesy Oil & Gas Journal)

WELL CONTROL AND SAFETY 165

and type of preventers assembled in the stack for a given well depend on the anticipated drilling conditions. A typical BOP stack might include an annular preventer on the top, a blind ram preventer below it, and a pipe ram preventer on the bottom. The stack is attached to the casing head, and the individual preventers in the stack may be separated by short lengths of pipe or spools for operational reasons.

Valves are usually provided on the side of one of these spools so drilling fluid returning from the hole can be regulated or shut off. Another set of valves allows fluid to be pumped into the well bore if it becomes necessary to kill the well.

Blowout preventers and related equipment in the BOP stack come in a range of pressure ratings. Each component must be selected to provide adequate strength for expected pressures at various depths in the hole. The BOP stack must also be arranged in the proper sequence.

BOP control. Blowout preventers are actuated (opened and closed) by a hydraulic system that is controlled from a console near the driller's position on the rig floor. The main control unit, however, is some distance away from the rig floor, and remote actuation of the BOP is possible if the rig floor is unsafe. The main control unit (Fig. 8–6)

Fig. 8-6 BOP control unit. (courtesy Hydril Mechanical Products Division)

includes accumulator tanks and pumps to pressurize the hydraulic lines that actuate the BOP elements and a manifold to direct the hydraulic fluid through the proper lines to accomplish the desired operation.

Choke manifolds. A choke manifold installed on the well site may play an important role in well control if a kick occurs.

A choke is a type of valve that resists flow and creates a pressure upstream of itself. Chokes are used in production operations as well as during drilling. They may be adjustable or equipped with a fixed-size orifice through which fluid flows.

The choke manifold used while drilling is a system of valves and piping that contains a choke through which flow from the annulus can be directed. It is connected to valves on the blowout preventer stack and is located a short distance away from the BOP stack. The choke and choke manifold are used when circulating a kick out of the hole and can direct flow to the mud pits and other mud handling equipment, to a reserve pit, or away from the well for burning.

Several types of chokes are available, including hand adjustable, and remotely controlled adjustable. Materials of construction vary, but chokes must be durable when used for drilling. The fluids flowing through the choke contain solids and other materials that can erode the choke or damage it in other ways.

BOP drills and training. The latest equipment for blowout prevention and well control is of little value if the crew members do not operate it properly at the right time. So BOP drills are conducted on a regular schedule on the rig to maintain the crew's proficiency in blowout prevention. Each member of the crew is assigned a position on the rig and a job to perform at that location. The smooth execution of each duty is critical when a blowout threatens.

Much emphasis has been put on well control training in recent years. Well control training is a key part of overall drilling training and is provided by a number of specialized schools around the world.

Early training of this type was done primarily in the classroom, and only a few training wells existed in which drilling conditions could be simulated. So the number of crewmembers that could be trained on these wells was limited. In recent years, the electronic training simulator has become widely used for training drilling crews in all phases of drilling operations, including well control. A variety of such simulators is available, many providing a realistic simulation of the drilling floor and conditions that exist on the rig while performing certain operations.

After the well blows out

Few standard procedures exist to follow if a well does blow out. Each situation is unique. Because of the variety of conditions, special equipment has been designed by firms offering well control services. In many cases, such equipment must be field-fabricated at the rig.

Some blowouts take months to bring under control and can cost tens of millions of dollars. Others may be brought under control in a few days. Some blowouts can be killed by operations performed in the well. In other cases, it is necessary to drill one or more relief wells to intersect the wellbore before control is possible.

Eventually, killing most blowouts depends on the ability to pump a heavy fluid into the hole—either in the original wellbore or through a relief well. Occasionally a blowout will bridge or stop flowing of its own accord, but such cases are rare.

Highly publicized examples. An incident in the Santa Barbara Channel offshore California in early 1969 probably did more to change the course of offshore drilling than any other nontechnical event.[3,4] Loss of control in a well being drilled from a fixed platform did little damage to equipment, and a relatively modest amount of oil was released into the ocean. But the oil spill's movement onto beaches in the area brought demands to halt offshore drilling and to enact strict rules governing offshore operations.

The incident also emphasized the difficulty in planning to cover any possible event that can occur while drilling.

On the Santa Barbara Channel platform, the drilling crew was pulling the drill string from the hole to prepare to run an electric log when the well began to kick. Kicks often occur while drilling, but with the proper equipment and properly trained crews they can be controlled before a blowout results. In the Santa Barbara incident, for example, the blowout preventer was reportedly closed to contain the kick. This action should have gained control of the well. But oil began to flow to the surface some distance from the wellbore, indicating that fluids were escaping from the hole at a point below the sea floor and were moving to the sea floor through a geologic fault. Such geologic conditions are not common, but there are ways to prevent such an occurrence even when such conditions do exist.

This incident emphasizes the wide range of possible events that must be considered in the well plan when preparing well-control procedures. It also indicates the consequences of even infrequent loss of well control. At the time of the Santa Barbara blowout, more than 1,100 holes had been drilled in the Santa Barbara Channel—some were producing wells, some were core holes—without a similar incident. Another report in 1971 noted that over 5,000 productive wells had been drilled on federal Outer Continental Shelf (OCS) leases since 1953 without an incident similar to the one in the Santa Barbara channel.[5]

Despite this record, that blowout virtually shut down drilling in the area for several years.

New regulations were soon developed governing drilling operations on the OCS areas involving casing, cement, mud, and blowout prevention equipment requirements; tests and inspections; safety and alarm equipment; and pollution control equipment.[6]

Unfortunately, other blowouts in the Gulf of Mexico occurred soon after the Santa Barbara accident and also received wide attention. In early 1970 a fire ignited on a platform in the Main Pass area offshore Louisiana (Fig. 8–7). That incident was not the direct result of drilling operations. But coming so soon after the California spill, it further influenced the tightening of controls on offshore drilling and producing

Fig. 8-7 Platform fire offshore Louisiana. (courtesy Oil & Gas Journal)

operations. Inspections of downhole safety valves in wells in the area, for example, indicated use of these devices had not been adequate, and stricter rules for downhole safety equipment resulted.

In late 1970, another fire erupted on an offshore platform in the Bay Marchand area in the Gulf of Mexico offshore Louisiana.[7,8] It was being fed by several wells on the platform. It was necessary to drill relief wells to reach the well bores contributing to the fire, and an oil spill resulted from the incident.

This particular blowout did not occur during drilling but apparently happened while a producing well was being repaired during a workover operation. When control of the well was lost and a fire ignited, surface control equipment on other wells on the platform was damaged. These wells then flowed out of control, contributing to the fire.

One of the largest oil spills from a drilling well resulted when the Ixtoc 1 well blew out in Mexico's Campeche Sound in mid-1979.[9,10] According to reports, drill pipe was being tripped out when gas began flowing from the hole and ignited. Oil was estimated to be flowing from the hole at a rate of 30,000 b/d, threatening to damage Gulf of Mexico beaches.

Many attempts were made to bring the well under control. At the same time, efforts to contain and recover the spilled oil from the ocean, including the use of some innovative techniques, were only partially successful. Relief wells to intersect the well bore so fluids could be pumped into the blowing well started shortly after the well blew out. But it was not until ten months after the incident occurred that the well was brought under control by two relief wells.

Though certainly not typical in magnitude, the Ixtoc well blowout indicates how costly an extremely serious blowout can be. In addition to tragic loss of life:

• While the well was out of control, lawsuits totaling nearly $400 million were filed in the U.S. against the two Mexican companies and one U.S. company involved in drilling the well.

• Oil from the blowout reached Texas beaches along a 140-mile span, prompting legal action by several groups.

• The operator of the well reportedly spent about $132 million to bring the well under control and contain the spilled oil.

• The well flowed an estimated 3 million bbl of oil before it was brought under control.

Bringing the well under control involved large amounts of special equipment, materials, and specially trained personnel for spill containment, relief well drilling, and other operations. An estimated 9–10 million bbl of water were pumped through the relief wells before the well was controlled.

Some perspective. These few incidents are by no means all of the loss-of-well-control situations that occur around the world. Nor is the seriousness or the consequences of those events discussed here typical of the bulk of wells in which control is lost. But their cost and their influence on offshore drilling and production operations emphasize the need for every effort to be made to prevent loss of well control.

Despite the number of blowouts that occur each year, the industry's safety record is remarkable. In the U.S. alone, over 80,000 wells were drilled in 1981, and only a very small fraction of those experienced serious loss of control. Considering the variety of conditions encountered—some unexpectedly—the record is good.

Also, the volume of oil spilled from offshore drilling and production operations is small compared with the oil spilled by tankers and released from other sources.

Increased efforts at well control have significantly improved the industry's record, but there is still a need for more advanced well control equipment and more well control training.

One study revealed that during the period 1971–1978, 46 blowouts occurred in U.S. Outer Continental Shelf waters, all in the Gulf of Mexico.[1] Of these, 17 happened while drilling exploratory wells from mobile offshore drilling rigs; 13 occurred during development drilling from fixed offshore platforms. The remainder involved production, workover, or completion operations.

Most of the 46 incidents "were of short duration and had minimal impact," according to the report by the U.S. Geological Survey. Of 17 exploratory well blowouts, 10 lasted from 15 minutes to one day; the rest lasted 21 days or less. All 13 of the development well blowouts were under control within 17 days; only one lasted more than one week.

Most of the blowouts during drilling were caused by the flow of overpressured gas at shallow depths. And during the period in which the 46 blowouts occurred, over 7,500 wells were begun on the U.S. Outer Continental Shelf.

The study indicates the industry's safety record is best for frontier-area drilling, apparently because special precautions are more frequently included in the well plan. Special care is also taken in drilling in deep water, the study showed. From 1975 to 1978, 67 exploratory wells were drilled on the OCS in water depths of 600 ft or more without a single blowout.

References

1. "OCS Blowout Record Seen Good, But . . . ." *Oil & Gas Journal*. (24 March 1980), p. 207.
2. Kennedy, John L. "Losing Control while Drilling: A 32-Well Look at Causes and Results." *Oil & Gas Journal*. (20 September 1971), p. 121.
3. "Huge Channel Oil Spill Blows Up Storm." *Oil & Gas Journal*. (10 February 1969), p. 50.
4. "Troublesome Channel Oil Leak Persists.' *Oil & Gas Journal*. (17 February 1969), p. 58.
5. "Environment Act Applied to Channel." *Oil & Gas Journal*. (1 March 1971), p. 32.
6. "Stiff New Rules Hit Channel Operators." *Oil & Gas Journal*. (31 March 1969), p. 36.
7. "Shell Snuffs Big Well Feeding Gulf Fire." *Oil & Gas Journal*. (4 January 1971). p. 37.
8. "Shell Gaining Ground on Big Gulf Fire." *Oil & Gas Journal*. (11 January 1971), p. 30.
9. "Pemex Recovers Half of Campeche Blowout Oil." *Oil & Gas Journal*. (11 June 1979), p. 33.
10. "Pemex Kills 1 Ixtoc Blowout after 10 Months." *Oil & Gas Journal*. (31 March 1980), p. 54.

9 Completing the Well

THE completion phase of an oil or gas well generally includes those operations that take place after the producing formation has been penetrated or the well has reached the planned total depth and before the well is put on production.

On many wells, completion operations are performed using the rig that drilled the well. It remains on the well until production equipment is installed. But on some wells, especially deep ones where a large expensive rig was required for drilling, the rig that drilled the well may be removed when actual drilling is done and a completion rig may be moved onto the well to perform completion operations.

These completion rigs, though often large for their type, are smaller than the deep drilling rig, cost much less to operate, and can be moved in and rigged up quickly. The functions of a rig may still be needed during the completion operation, particularly hoisting, but the weights to be handled are much less than the weights of drill pipe and casing that must be hoisted during drilling. The smaller rig can perform these completion functions adequately.

The time required for completion work is a factor in whether the drilling rig or a smaller unit will be used. A routine well completion that can be done quickly would not warrant the inconvenience and delay of moving one rig out and another one in. But if extensive testing and other work will be done over a long period, moving in another rig is easily justified by the lower cost of the completion rig.

Production equipment is installed during the completion phase. But before that is done, other operations must be conducted in most wells. These include the following:

1. logging the hole to measure formation characteristics and to help determine the producing capability of the pay zone

2. drill-stem testing, using the drill string and special test equipment that is lowered in the hole
3. stimulation or treatment of the producing zone to increase its flow capacity
4. perforating the production casing to allow formation fluids to flow into the well bore

Logging and drill-stem testing may be done before the production casing is run. Tools are available with which to perform both operations in either an open hole or inside casing. But it is often not evident until after logging or drill stem testing whether the formation will be productive enough to justify completing the well—whether the well will be commercial.

If the well can be completed as a producer, production casing is run and cemented. If the completion is to be a cased-hole completion, casing is run to a point below the producing formation. If an open-hole completion will be made, the casing is stopped above the producing zone. Both types of completions are used, but the cased hole is more common.

If a cased-hole completion is made, the producing zone is isolated behind the production casing. Then the production string must be perforated so fluids can enter the well bore and flow or be lifted to the surface. In an open-hole completion, it may be necessary to use one of several techniques to prevent the wall of the hole through the producing zone from sloughing into the well-bore as fluids are produced.

Logging

All of the evaluation techniques used in assessing formation characteristics and well potential are best applied when considered together. Especially in exploratory wells, no one type of log—or even several logs without the information provided by a drill-stem test—will evaluate the well adequately.

Logging and testing are aimed at measuring the porosity, permeability, and fluid saturation of the reservoir rock—key properties that determine the well's producing capability.

Porosity, as it applies to oil and gas reservoirs, is the amount of space in the rock that is available for the storage of hydrocarbons. It is the ratio of void space in the rock to the rock's bulk volume, multiplied by 100 to express it as a percentage.

Permeability is the ability of the formation to conduct fluids or the lack of resistance to the flow of fluids. The unit of permeability is the darcy.

Reservoirs normally contain both hydrocarbons and water. Fluid saturation is the quantity of oil, water, and gas that is contained in the formation rock.

It is possible to have relatively good porosity and relatively low permeability; such a formation would have ample pore space to contain oil or gas, but the fluids would not flow easily from these pore spaces into the well bore. And even though porosity would be relatively high, it might be that those pores would contain mostly water rather than oil or gas.

Types of logs. The first log of the type used in oil and gas well drilling was an electric log, first run in France in 1927. It measured the resistivity of the formations at a number of points in a well bore. Then the measurements were hand-plotted on a graph.

Electric logging was soon introduced in the U.S. In the more than 50 years since then, logging of oil and gas wells has developed to a highly sophisticated technology that permits measuring a variety of reservoir rock characteristics, plotting them continuously and automatically, and analyzing the measurements by computer.

The first logs measured only the electric resistivity of the formations, or their electric self-potential. Today's logs can measure these properties plus natural and induced radioactivity, acoustic properties, and other characteristics. Often a number of different types of logs—a suite of logs—is run on a single well to determine rock and fluid properties more accurately. Excerpts from several well logs are shown in Fig. 9–1.

The value of well logs, however, lies in proper interpretation. A highly specialized science, log interpretation is the key to accurate reservoir evaluation, which in turn leads to the most efficient exploitation of oil and gas reservoirs.

In electrical logging methods, clean sands, sandstones, oil, and gas offer a high resistance to the flow of electric current; they act as insulators. But water is very conductive, offering little resistance to the flow of electric current. Measuring the resistivity, then, indicates the type of formation and the type of fluids, if any, in the formation rock.

Softer formations—relatively unconsolidated sand and shale sequences, for instance—have lower resistivities. Hard formations such as limestones and dolomites have higher resistivities. Impervious formations such as anhydrite have extremely high resistivities because their fluid content is low.

The self-potential or spontaneous potential (SP) log measures the

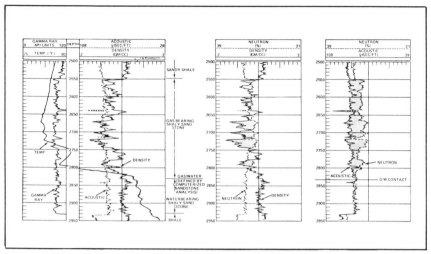

Fig. 9-1 Portion of well log. (courtesy Oil & Gas Journal)

electrical potential in the mud around the logging tool. It is not related directly to the porosity and permeability of the formations, but it is useful in correlating formations between wells in a given area and in making a qualitative analysis of a zone to indicate where other logging tools should be run for a more detailed study. Other types of logs are usually run in conjunction with the SP log.

Formation delineation with electric logs is often difficult when saltwater or oil muds are being used, and it is sometimes necessary to use radioactivity logging methods. Shales and clays exhibit a high natural radioactivity; sandstones have radioactivity only if they contain other minerals in the matrix; and limestones and dolomites usually are low in radioactive elements.

The gamma-ray and neutron log are usually run together, the gamma-ray log indicating which formations may be porous and the neutron log showing which rocks are porous and permeable.

The gamma-ray log is used to differentiate between shales and other formation types and can be used in cased holes to indicate the shale content of formations. The neutron log measures the relative amount of fluid surrounding the logging tool and delineates porous formations. It measures the hydrogen content of the fluid; increased hydrogen content indicates a greater amount of fluid and pore space. These radioactive logs cannot be used to differentiate between oil and water, so a resistivity log must also be run for this purpose.

The formation density log uses a gamma-ray source and a detector.

The gamma rays are scattered. Those that reach the detector are measured, indicating the formation density.

An acoustic velocity log (sonic log) measures porosity between about 5% and 30% and is often used to distinguish between salt and anhydrite. The sonic log must be run in an uncased hole. It does not measure permeability. Shaliness or other dirt in the formation will affect the transit time of the sound waves and, hence, porosity calculations.

These are the common types of logging tools, but there are also special logs for special purposes. The combination of these logs that is run on any given well varies with the type of formation, the type of drilling fluid in the hole, whether the hole is cased, and other factors.

Coring. Cores taken from a well bore offer a considerable advantage in formation analysis. They are used for a detailed examination of the types of rock penetrated. They can provide a basis for calibrating logging tools, and they give the best porosity and permeability data possible because the actual rock from the formation can be tested in the laboratory.[1] Then wire-line logging and drill-stem test results are compared with the results of core analysis.

Coring is not done on all wells; it is more applicable in wells where little information is available from previous drilling or testing. It would be unlikely, for instance, to core in a development well that was one of many already drilled in a field. It would be more likely to core in an exploratory well where less information was available.

Coring is done with a three-cone roller coring tool, a diamond cutting tool, or a wire-line tool that can be used with either the roller bits or diamond bits. In one method, the core is retrieved by pulling the core barrel containing the core to the surface with the drill pipe. In the wire-line technique, the core is retrieved by lowering a wire line with a tool that attaches to the core barrel so the core barrel is retrieved without pulling the drill pipe.

The core barrel, which receives the cylindrical core as it is cut by the bit, is normally made up in 30-ft lengths. For example, 30, 60, or 90 ft of core can be cut at one time. The wire-line core barrel is normally 15 ft long.

Another type of coring device is a side-wall coring tool. It is lowered to a predetermined depth and bullets are fired electrically to force small cylinders into the wall of the hole to extract plugs of the formation rock. These samples are then retrieved to the surface for analysis. Since the samples are small, they are not adequate for making precise porosity and permeability determinations. They are used only to study a specific formation to verify the type of rock and to see what type of fluids exist in the formation.

Side-wall coring is expensive and is used primarily in exploration wells—but certainly not in all of them.

With all coring operations, the hole must be clean before coring begins. Any junk must be removed from the hole, by fishing if necessary, and the hole must be conditioned by circulating drilling fluid for a considerable period of time before beginning to core. It is also important that the bit be kept clean while coring and that the proper bottom-hole assembly be run to stabilize the lower portion of the drill string.

Analysis of conventional cores (not side-wall cores) is done using a porometer to measure porosity, a permeameter to measure permeability, and a saturation retort to measure fluid saturation.

Drill-stem testing

After the prospective formation has been penetrated, drill-stem testing is often done before the production casing is run. It is perhaps the most important indicator of whether a formation will be productive.

Drill-stem testing offers a way to relieve the hydrostatic pressure of the drilling fluid on the formation to be tested without removing the drilling fluid from the hole (Fig. 9–2). The drill string is run in the hole with a tail pipe or anchor pipe on bottom through which fluids from the formation flow into the drill string and to the surface. Above this section of pipe is a packer that is actuated when the drill string reaches bottom. The packer isolates the column of drilling fluid from the formation to be tested. When the packer is set against the wall of the hole or the casing and the drill string is opened to the surface, pressure is relieved from the formation and fluids can flow up the drill pipe to the surface.

Drill-stem tests can be run in either an open hole or in a cased hole, using the appropriate packer and other equipment. Two packers can also be run on the drill string to isolate the formation of interest from adjacent formations.

A device in the tool records pressures throughout the test period. The recording chart is driven by a clock so the pressure reading can be correlated to any time during the test. The measurements of primary interest in analyzing the drill-stem test, using the pressures recorded on the down-hole recorder, are initial hydrostatic pressure, initial flow pressure, initial shut-in pressure, final shut-in pressure, final flow pressure, and final hydrostatic pressure.

By analyzing the pressure record, much can be determined about the

178 FUNDAMENTALS OF DRILLING

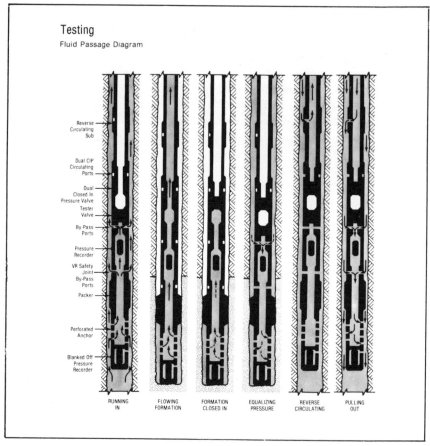

Fig. 9-2 Drill-stem testing. (courtesy Halliburton Services)

producing capability of the formation. The drill-stem test can serve as a preliminary reservoir evaluation test and as an indicator of whether the well is commercial and completion is justified.

Perforating

If the well is judged commercial based on log analysis, drill-stem test results, and other information, the production casing is run. The production string provides a stable hole through the producing interval, and producing equipment will be installed in the production string. But access to the inside of the production string must exist so reservoir fluids can enter the well bore.

To provide this access, the casing is perforated in the producing interval. Perforating involves lowering a special tool to a predetermined depth inside the production casing then activating small explosive charges that make holes in the casing.

A firm specializing in this operation comes to the well site to perform the perforating; the job is not done by the drilling contractor. Equipment includes a perforating truck equipped with a cable drum containing the amount of cable required for the particular depth at which perforating will be done, and sophisticated instrumentation that records the depth of the perforating tool and activates the charges when the tool is in position.

The perforating guns can be lengths of cylindrical steel with evenly spaced holes through which the charges are directed when activated. Other types of perforating guns use charge carriers that are not contained in a cylindrical housing.

There is a practical limit to how long a perforating gun can be—how many sections can be made up for one trip in the hole. But the length can be varied according to how long an interval must be perforated. One of the factors that limits the length of a perforating gun is the amount of hole deviation. Since the gun is run on a wire line rather than on drill pipe, very little weight is available to force the gun through any tight spots in the hole.

If a long interval is to be perforated, more than one perforating trip in the hole may be necessary. The portion of the interval that will be perforated on each trip is determined, and on each trip the tool must be lowered to the proper depth before the charges are activated.

Considerable research has gone into determining how a particular gun and charge design perform. It is desirable to penetrate as deeply as possible through the casing and into the formation. Different types of guns and charges have been developed for different conditions. In a single completion—a well producing from only one reservoir—it is not extremely important which side of the casing is perforated. The depth at which the perforations are made is very important, but the direction in a horizontal plane through the well bore is less critical. However, many wells are multiple completions—more than one zone is produced in a single well bore. It is sometimes necessary to perforate these several zones. Then the direction in the horizontal plane in which the gun is discharged becomes extremely important.

For instance, in perforating the uppermost zone in a multiple completion, the tubing string for a lower zone may be in place opposite the point at which the casing must be perforated. In this case, oriented perforating is done. A radioactive source can be lowered into the tubing that reaches to the lower formation. Then the perforating gun is lowered to the depth required for perforating the casing opposite the upper zone. When the perforating gun is at the proper depth, it is rotated by controls on the surface and the response of a recorder to the radioactive source in the other tubing string is measured at the surface. A plot is made of the relative position of the charges in the perforating tool to the tubing string containing the source material. Then the perforating tool is rotated so the charges will be directed away from the tubing before the gun is fired.

Sand consolidation

In many wells, it is necessary to provide some means for preventing solid material in the formation from falling into the well bore and filling up the bottom of the hole or from flowing into the well bore with reservoir fluids and to the surface. For instance, sand entering the well bore can erode and plug pumps and other equipment, both downhole and at the surface. To prevent problems associated with solid formation materials entering the well bore, several techniques are available.

Gravel packing is one common method of controlling sand present in a producing formation. It is normally used in relatively shallow wells where formations are unconsolidated. In one example after the hole was drilled past the producing formation, a reamer bit was run to enlarge the diameter of the hole in the producing interval to receive the gravel pack.[2] Then a slotted section of casing was run and placed opposite the producing zone. The casing string was fitted with casing packers at appropriate intervals to isolate zones. Ports in the casing at the top of the zone to be gravel packed let the gravel packing material flow out of the casing and behind it into the enlarged hole section. With a packer set below the section to be gravel packed, fluid used to carry the gravel packing material to the producing interval returned through slots in the slotted casing section, to the tubing inside the casing, and back up the tubing to the surface.

Packing material can also be placed through perforations in the casing.

In all sand consolidation techniques, the carrier fluid used to place the packing material in the well bore must be chosen carefully. A variety of fluids is used for this purpose, some specially designed to provide specific properties required for suspending and placing the pack material.

Stimulation

Two general types of well treatment or stimulation are common. Hydraulic fracturing is done to fracture the formation and create channels in the rock through which fluids can flow to the well bore. The fractures also increase the area from which formation fluids can flow from the rock.

Fracturing is done by pumping a liquid into the formation at pressures high enough to crack the rock (Fig. 9-3). This pressure is determined from the fracture gradient information discussed in chapter 6. A solid material—normally sand, but other materials have been used, including bauxite—is pumped into the formation along with the fracturing fluid to prop open the fractures after they are created. When the pressure applied to create the fractures is released, the fluid flows back into the well bore and the sand (the proppant) remains in the fractures.

The other common treatment of producing zones is acidizing. This technique increases productivity by dissolving soluble material from the formation rock, allowing the exposure of more formation area. A

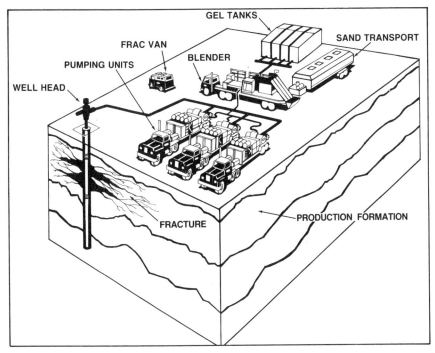

Fig. 9-3 Hydraulic fracturing equipment on location. (courtesy Halliburton Services)

fluid is pumped down the well into the formation. It is typically a mixture of 85% water and 15% hydrochloric acid (HCl), although other acids are also used.

For a conventional acid treatment, required pressures are not as high as those needed for a fracture stimulation because the formation being acidized does not need to be fractured. There is, however, an acid frac treatment that combines the two techniques for treating formations with special characteristics.

Not all wells must be stimulated before they are put on production. Many producing wells are acidized or fractured after they have been producing for some time to increase productivity or to return productivity to initial levels. Some wells are stimulated with one of these techniques several times during their producing lives.

Whether or not a newly drilled well requires stimulation before it is put on production depends primarily on the type of formation. Another important factor is whether the producing zone was damaged during drilling—for instance, by the drilling fluid. A thick filter cake on the wall of the hole through the producing zone can inhibit the flow of formation fluids into the wellbore. If possible, it is best to use as light a drilling fluid as possible when drilling near the prospective producing zone. The properties of the fluid must also be adjusted to minimize the chance of reducing the formation's productivity. One of the advantages of oil-base drilling fluids is that they generally cause less long-term damage to the producing zone.

It is not always easy to determine how much, if any, formation damage has occurred at this point. After the well is put on production, it may be necessary to perform more extensive flow tests to determine the damage. Methods exist for estimating the theoretical flow capacity of the formation and for comparing it with the apparent flow capacity. If significant damage is indicated, it may be necessary to treat the well further after it has been on production.

Producing equipment

Most wells are equipped with tubing through which production from the well flows to the surface. Tubing is installed inside the production casing and may be equipped in various ways. In most cases, a packer is installed in the tubing string above the producing formation to isolate the pay zone from the rest of the hole and to relieve the pressure exerted by the fluid between the tubing and casing.

In multiple completions, special packers can accommodate more than one string of tubing, so an individual tubing string can be installed for each producing zone. This allows flow from each zone to be isolated. In such a well, the upper packer isolates the hole and fluids above the top producing zone and lets the tubing for the bottom producing zone pass through the packer. Below the upper producing zone, another single packer isolates the lower zone from the upper zone. Formation fluids enter the well bore in each case below the respective packer and flow into the tubing that is open opposite that zone. Fluids from the upper zone cannot enter the tubing that is open opposite the lower zone, and vice versa.

Pumps. Gas wells flow unaided through the tubing to the surface where water, acid gas, and gas liquids are removed by surface equipment at the well or in a gas-processing plant. Some oil producing wells also have enough reservoir energy so fluids will be forced to the surface. In these wells, oil flows up through the tubing to the surface by the natural force of the reservoir.

But in many reservoirs, only enough natural pressure exists to lift the fluid partway up the well. Then the hydrostatic pressure of the fluid column equals the natural reservoir pressure. In these wells, some means of artificial lift is used to supplement the reservoir's natural energy.

Artificial lift techniques include rod pumping, subsurface hydraulic pumping, gas lift, and submersible pumps. The most common, rod pumping, is a mechanical system that includes a surface pump jack driven by an electric motor or gas engine (Fig. 9–4). The reciprocating motion of this beam pump, or horsehead pump, at the surface moves a string of rods vertically. The rods are connected to a downhole pump installed in the tubing near the producing formation. This pump consists of a barrel containing two check, or one-way, valves.

On the upstroke of the rods, the upper valve closes and the plunger lifts the oil above that valve, raising the level of the oil column in the tubing. At the same time, because the hydrostatic pressure is relieved inside the pump barrel by the upward motion of the pump plunger, the lower valve opens, allowing more formation fluid to flow into the pump barrel. On the pump's downstroke, the upper valve opens, allowing the plunger to move down through the oil in the pump barrel. At the same time, the lower valve is closed by the hydrostatic pressure of the fluid column, preventing oil from flowing back into the formation.

This pumping method is the most common. Not all such systems are powered by the traditional horse-head surface unit, however. Other

184 FUNDAMENTALS OF DRILLING

Fig. 9-4 Beam pumping unit. (courtesy Lufkin Industries Inc.)

pumping systems using rods and a downhole plunger pump have been designed in an attempt to increase efficiency and to reduce stresses on the rod string.

In subsurface hydraulic pumping (Fig. 9-5), oil is pumped down the well to drive a hydraulic motor, which in turn drives a pump to move oil to the surface. Oil produced from the field is used as the driving force for the downhole pump. This downhole equipment is also installed in the production casing.

Another approach to artificial lift is the use of submersible pumps. The submersible pump (Fig. 9-6) is a multistage electrical centrifugal pump that operates completely submerged in the fluid it pumps. It is connected to a surface power source by electrical cable and is run into the hole on tubing inside the production casing. Major components are the pump, the motor, a seal assembly, the cable to the surface, and the necessary surface controls.[3] Such pumps can be used on both onshore and offshore wells. The most common applications are wells that produce large amounts of water with the oil, wells that produce highly

COMPLETING THE WELL 185

Fig. 9-5 Subsurface hydraulic pumping system. (courtesy Guiberson Division, Dresser Industries Inc.)

viscous fluids, and wells that produce oil which contains a large amount of gas.

Yet another approach to oil production is the use of a gas-lift system. This artificial lift method normally is most applicable in fields or areas where gas is available. Gas is injected into the tubing at intervals in the well bore to lighten the weight of the fluid column and to reduce the hydrostatic pressure of the column on the producing formation. Valves that permit the gas to enter the tubing are set to open and close at predetermined pressures exerted by the fluid column. These valves are installed on the tubing as it is lowered into the hole. To supply the gas-lift gas at the pressure required, surface compression facilities are usually required.

The importance of artificial lift methods in oil production, especially in the U.S., is evident from a study made in the late 1970s which indicated that, of the more than 516,000 onshore oil producing wells in the U.S., only about 12% flow without arificial lift.[4] Of the remaining

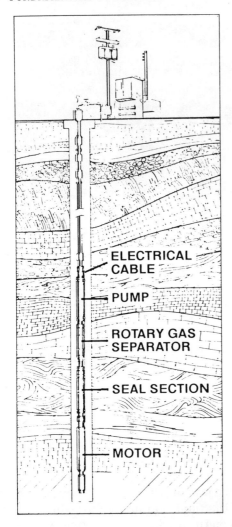

Fig. 9-6 Submersible pump.
(courtesy Centrilift Hughes)

wells that are artificially lifted, over 386,000 are produced by rod-pump systems, about 48,000 are produced by gas lift, nearly 9,500 are produced by subsurface electric submersible pumps, and roughly 8,600 are produced by subsurface hydraulic pumping systems. Outside the U.S., wells are generally more productive, although capacity varies widely. The result is fewer wells, and a smaller portion of those require artificial lift.

Other producing equipment. Besides the tubing, pumps, and related equipment installed in the well during completion, other equipment may also be needed.

Downhole safety valves are one example. There are several types of downhole safety valves. Some are direct-operated with no connection to the surface; others are remotely controlled from surface actuators. Direct-controlled subsurface safety valves are set to close at a predetermined pressure or at an abnormally high flow rate. For instance, if a line break or other loss of well control at the surface results in unrestricted flow from the well, the subsurface safety valve will close automatically, shutting off the flow of fluids. Surface-controlled subsurface safety valves are held open by hydraulic pressure in a control line and are much easier to include in an automated well control system.

A number of regulations and standards govern the use of safety valves both in wells and on other producing equipment. The most stringent of these apply to offshore producing facilities where much stricter rules resulted, in part, from the serious oil spills of a decade ago.

Testing after completion

After the well is completed and all producing equipment is installed, more testing may be required. Further treatment or stimulation of the producing zone may also be needed at any time after completion. This is particularly true if production tests indicate the formation was damaged by the drilling fluid or by drilling operations.

Soon after completion, a potential test is usually performed to determine the producing capability of the well. The well may be produced through either temporary or permanent equipment during this test, including a separator to separate gas, water, and oil in the well stream and storage tanks to receive the produced oil.

Important test data include the flow rate, bottom-hole pressure in the well while flowing, pressure after the well is shut in, and the amount of gas, oil, and water produced. Analysis of this information helps determine the producing capacity of the well—its potential—as it was completed. It also can indicate if damage to the producing zone occurred during drilling or completion.

Reservoir engineering procedures can be used to estimate the producing capacity of the well if no damage was incurred. This theoretical producing capacity is compared with the actual producing capacity indicated by flow tests to determine if formation damage exists. If damage is suspected, it may be desirable to treat the well again or to modify the original well completion to improve productivity.

The productivity of an oil well can be determined by first measuring the bottom-hole pressure while the well is shut in and then measuring bottom-hole pressure at several oil producing rates.[5] After measuring the shut-in bottom-hole pressure, the well is opened at a low flow rate and is allowed to flow until the bottom-hole flowing pressure stabilizes. The flow is then increased, and pressure and rate are again recorded. This sequence is repeated for several flow rates. Pressure differences are then plotted against flow rate, and a productivity index is determined, given in units of barrels per day per pounds-per-square-inch.

Potential tests are also conducted on gas wells after the well is completed. One approach to estimating the producing capacity of a gas well is the four-point or calculated absolute open flow (CAOF) test. The test is performed by flowing the well at four different rates and measuring flow and pressure data at each rate. By using reservoir engineering methods to analyze the test data, the operator can determine the calculated absolute open flow potential of the well. Part of the procedure is to plot on graph paper the pressure and flow volume at each rate, hence the term four-point test.

The CAOF is, as its name implies, a calculated value, and the well is not produced at the calculated absolute open flow rate. The CAOF is useful in comparing the relative potential of different wells, but it is only a calculated value. For example, during a four-point test the well might be produced at rates of 2, 4, 6, and 8 MMcfd. But the calculated absolute open flow, depending on the reservoir pressure, might be 30 MMcfd.

A four-point test may take two or three days, depending on how quickly pressures in the well stabilize at each flow rate. Other tests may take much longer, depending on the objective of the test. In a new field, for instance, it may be desirable to flow a well, or several wells, for extended periods to help define the areal extent of the producing zone. If a well flows for a very short time, the effects of that flow on reservoir pressure and producing capacity extend only a short distance from the well bore. The longer the well is produced, the farther from the well bore the effects will reach. By flowing the well for an extended period and monitoring pressures, the long-term producing capacity of the producing zone can be analyzed. If several wells have been drilled that can be monitored during the test, the analysis becomes more precise.

Such long tests are more applicable in newly discovered fields with only a few wells. In an established, mature field where much is known about the reservoir, such extensive testing is usually not required.

A well is tested periodically during its life, both to satisfy regulatory requirements and to provide the owner with operating information that will help determine if the well and reservoir are being produced efficiently. If several wells produce into a common facility in a field, separate test equipment is normally installed. It isolates an individual well's production for testing. In an oil production facility, a test separator is usually provided. The well being tested flows through that separator while the other wells are processed together through the production separator. Flow rate data, water and gas content, and other information are recorded during the test.

Throughout the producing life of an oil or gas well, the operator monitors well performance and attempts to obtain the most efficient recovery of oil or gas that can be obtained from that well and from the field. Analyzing these tests over a period of time can indicate if the well's productive capacity is deteriorating or if that well is producing increasing amounts of water or gas. If production declines, well treatment may be needed. It may be necessary to change the downhole equipment so production comes from a different interval of the pay zone, or it may be necessary to install additional equipment, including an artificial lift system.

References

1. Anderson, Gene. *Coring and Core Analysis Handbook.* Tulsa: PennWell Publishing Co., 1975.
2. Parmigiano, John M. "Multizone Gravel-Pack Completion Method Works in High Angle Holes." *Oil & Gas Journal.* (12 January 1976), p. 97.
3. Legg, Leo V. "Simple Plan For High-GOR Wells Improves Efficiency." *Oil & Gas Journal.* (9 July 1979), p. 127.
4. Fincher, L., and F.D. Griffin. "Rod Pumps On Offshore Platforms Lift Emeraude Field Production." *Oil & Gas Journal.* (3 November 1980), p. 68.
5. Amyx, James W., Daniel M. Bass Jr., and Robert L. Whiting. *Petroleum Reservoir Engineering.* New York: McGraw-Hill Book Co., 1960.

10 Today's and Tomorrow's Technology

THE conventional rotary drilling method that is used to do the bulk of the oil and gas well drilling around the world today will continue to dominate for the foreseeable future. Most drilling systems will still be built around the concept of rotating the bit on bottom and circulating a drilling fluid.

Equipment and techniques will undergo steady improvement, and there are some important trends developing in drilling tools and methods. Significant new systems and components are already out of the laboratory and being put to commercial use. Existing tools are being used in a wider range of applications.

During the 1980s new directions in drilling technology are expected. Many ideas will certainly develop, but the following examples of emerging technology will receive industry emphasis:

1. new bit design concepts promise to take a sizable share of the market in the next few years
2. rapid expansion of the use of measurement-while-drilling techniques will give the driller and drilling engineer more precise, real-time drilling data
3. greater use of improved downhole motors will reduce problems associated with high drill pipe rotational speeds and will provide an economic alternative for straight-hole drilling as well as directional work
4. development of drilling fluids and other materials and techniques capable of withstanding the extreme temperatures and pressures involved in deep drilling will continue to get a share of the industry's research and development time and money

5. rig crew training, using sophisticated equipment such as electronic simulators, will receive a high priority
6. instrumentation will become more sophisticated and more useful, and computer analysis of drilling data will improve efficiency

The fact that today's basic drilling techniques will continue to do the bulk of oil and gas well drilling does not mean the search for different ways to drill will be abandoned. But most of industry's time and money will be spent on less-than-revolutionary changes in conventional rotary drilling.

New bits

The common types of bits—milled tooth, insert, and conventional diamond—will also continue to be used for much of the world's oil and gas well drilling. But in the early 1980s, a bit that is significantly different from the traditional rolling cone bit began to be used increasingly.

A new approach. Polycrystalline diamond compact (PDC) bits promise to provide a faster, more economical way to drill in many formations. They remove rock by shearing rather than by crushing or grinding as rolling cone bits do.

The bits have a body into which drill blanks are inserted. The drill blanks consist of a layer of polycrystalline synthetic diamond on a tungsten-carbide substrate and are formed using high temperature and pressure. Inserts are arranged in a variety of patterns in the bit body, depending on the manufacturer.

The cutting structure and drill blank were developed by General Electric in the 1970s and are trade-named Stratapax. A number of companies manufacture the new type of bit, using the Stratapax drill blanks.

PDC bits have shown a potential for greatly extending the time a single bit can be left on bottom drilling and for providing good penetration rates. The longer bit life means fewer trips to change bits, reducing well cost.

Trip time and bit life are not the only variables in the drilling cost equation; bit cost is also an important factor. But in moderate to deep drilling, the longer bit life can make a more expensive bit economical.

Since the polycrystalline diamond compact bit does not remove rock by crushing or grinding, penetration rate is less dependent on bit weight than is the case with rolling cone bits. Therefore, considerably less bit weight can be used with the PDC bit. By drilling with reduced bit weights, the tendency of the hole to deviate is reduced. In addition, the design of the face of the bit may be a factor in reducing hole deviation tendency, according to one manufacturer.[1] Higher rotating speeds can also be used, making the bit compatible with downhole drilling motors.

Increased penetration rates and savings in trip time have been significant in many cases. A PDC bit with a turbodrill increased penetration rate fourfold over that obtained in similar wells drilled with conventional bits.[2] One PDC bit was used to drill more than 3,000 ft of hole; six conventional rock bits had been required to drill a similar interval.

PDC bits are much better suited for some formations than for others. Hard sandstones, for instance, seem to resist the cutting action of PDC bits. Shales, claystones, and siltstones, however, seem to drill well with them.

Since these bits shear or peel off the formation, it is important that the drilling fluid and hydraulics program be designed to prevent formation material from sticking to the bit. Oil mud seems to have been effective in this regard. In the North Sea, it has been used as the drilling fluid in most successful PDC bit runs.[2]

The industry has developed rolling cone and conventional diamond bits to a high degree of technology through many years of field experience and research. Large amounts of field and lab data have been used to determine the best bit weights, rotary speeds, and hydraulic conditions for each type of bit in a specific formation. Because of this maturity, it is unlikely that the industry can further develop traditional bits to achieve a drastic increase in penetration rate or efficiency.

The PDC bit, made possible by new materials, is still in an early stage of development. To date, most research has been aimed at developing the cutting material and methods for attaching the compacts, or drill blanks, to the bit body. Much more work will be needed to determine the best bit body designs, the optimum arrangement of the drill blanks on the body, the design and arrangement of jets for efficient cleaning, and the optimum bit weights, rotary speeds, and drilling fluid hydraulics.

But the early success of the basic concept indicates that use of these bits will expand rapidly. As more such bits are used in the field, the library of field data will grow, allowing improvements in design and optimization of operating conditions similar to those that brought rolling cone and conventional diamond bits to their current advanced state of technology.

Other bit types. There will undoubtedly be improvements in the traditional bit types. They certainly will not be displaced immediately by PDC bits.

For instance, extended-nozzle rolling cone bits have undergone considerable field testing. The combination of extended nozzles and a center jet has shown penetration rates can be increased from 15 to 40%.[2] Other work has centered on combining longer nozzles with a special nozzle design that causes cavitation, or the formation of areas of low pressure, as the drilling fluid exits the bit nozzle.[3]

As discussed in chapter 5, drilling is often done with a weighted drilling fluid that provides a substantial overbalance of the formation pressure in order to prevent blowouts. One result of this overbalance at the bottom of the hole is that the rock chips must be lifted against this pressure differential. The less the overbalance, the less energy is needed to remove the rock and the faster the penetration rate.

According to a report by Johnson et al., "The erosion and cleaning effect of jets is enhanced when the degree of cavitation occurring on or near the bottom of the hole is increased." Nozzle designs studied in this project could be suitable for existing mechanical drill bits.

Measurement while drilling

Two problems have existed with conventional means of measuring bottom-hole conditions and using those data to drill efficiently and safely: 1) bottom-hole conditions must be extrapolated from measurements made at the surface, and 2) there is a time lag between an occurrence at the bit and its indication by surface measuring devices.

For many years, the industry has sought to develop equipment and methods to overcome these obstacles to more timely, more accurate drilling data. In the late 1970s, work began to accelerate as more than 40 companies were involved in the development of such systems.[4]

Measurement while drilling (MWD) or borehole telemetry describes systems that measure drilling factors near the bottom of the hole and transmit the resulting data to the surface immediately. These measurements can be made while circulating or drilling, and the information provided is continuous. The need to stop drilling to run separate survey tools can be eliminated in many cases.

By giving the drilling engineer real-time data from the bottom of the hole without the need to extrapolate bottom-hole conditions from surface measurements, drilling safety and efficiency can be improved significantly. Further development of such systems could even provide the capability to look ahead of the bit and predict what type of formation will be encountered next. In addition to improving drilling safety and efficiency—well control, directional well surveying and control, and drilling optimization—these systems could provide real-time logging and formation evaluation.

The advantages that MWD systems offer are particularly attractive in offshore drilling where development wells must be drilled directionally. Control of the path of offshore wells is extremely important, and eliminating the need to stop drilling and survey well bore direction, for instance, can reduce drilling costs.

Improving drilling efficiency by optimizing drilling variables could be the most significant cost-saving feature of MWD systems. But more efficient drilling resulting from data supplied by MWD systems must be obtained in conjunction with adequate well control because conditions that allow faster drilling may adversely affect well control.

Formation evaluation measurements provided by MWD systems are not expected to replace conventional logging and other formation evaluation techniques. The sensitivity of MWD systems will not be adequate to provide the detailed information that can be obtained with today's sophisticated logging systems. But the formation data that MWD systems could provide would be of value to the drilling engineer as the well is being drilled. Also, preliminary information provided by an MWD system could help select the most appropriate log, or suite of logs, to be run.

Cost savings. Directional drilling is the most immediate and most obvious area in which real-time measurements from the bottom of the hole can pay off. Offshore development drilling from fixed platforms is a much larger potential application for such systems than onshore directional drilling is.

An important factor in assessing the potential savings from use of these systems in offshore drilling is that offshore platform drilling costs are typically twice the cost of an onshore well of equal depth. Offshore development wells must often be highly deviated since 18–36 wells are typically drilled from the same platform. While drilling a 9,000–10,000-ft offshore development well, about 60 directional surveys might be made, consuming 60–80 hours of rig time.

The average platform well takes about 30 days to drill and, assuming a cost of $20,000/day, total cost would be $600,000.[5] It is estimated that a telemetry system that eliminates the need for conventional directional surveying can save 2½ days time ($50,000) or about 10% of the well cost.

Improvements in drilling efficiency resulting from the use of MWD systems are more difficult to estimate. Projections of overall improvement in drilling efficiency have ranged upward to 25–30%, but this may be optimistic. As much as a 15% improvement in drilling efficiency might be possible, however.

How the systems work. Of the components in an MWD system, the method of transmitting signals from the bottom of the hole to surface recorders has given developers the biggest challenge. The system must be able to transmit signals at a reasonable speed from sensors housed in a drilling sub at the bottom of the hole to a surface unit that processes and displays the data.

Most MWD systems are included in these four types: mud pressure pulse, acoustic, hardwire, or electromagnetic. Following is a brief description of how the systems work and the advantages and disadvantages of each.[6]

The mud pressure pulse system transmits coded signals by restricting mud flow inside the drill pipe and creating an increase in standpipe pressure. These pulses from downhole sensors are decoded at the surface. Advantages of this system are its low cost and ease of handling; disadvantages are a relatively slow data transmission rate and possible excessive signal attenuation.

Acoustic methods transmit sound waves through the drill pipe or through the mud stream. The scheme is similar to the mud pulse system and uses the same general type of rig equipment. Acoustic methods are simple and low cost, and they have a higher data transmission rate than mud-pulse systems; signal interference is a possible disadvantage.

Hardwire systems require special drill pipe or inside-the-string tools with or without a jumper cable. Advantages of this type of system are the possiblity of two-way power and data transmission and very high data transmission rates. Disadvantages of hardwire systems include high cost, problems in handling the special drill pipe and connectors, and difficulty in providing reliable electrical connections.

Electromagnetic methods transmit signals through the earth and the drill pipe and have the potential to provide a high data transmission rate. These systems are relatively low in cost and do not present handling problems on the rig.

One commercial mud pressure pulse MWD system for directional data transmits measurements of inclination, azimuth, and tool facing from the bottom of a deviated well to the surface.[7] The downhole assembly is contained in a special 34-ft drill collar available in diameters of 7¾ in., 8¼ in., and 9½ in. Electrical power is provided downhole by a mud-driven turbine generator.

This tool (Fig. 10–1) is used primarily offshore in drilling directional wells from kickoff to target depth. It can be used in conventional rotary drilling, while drilling with mud motors or turbines, or when jetting. According to the manufacturer, the system can take a directional survey in less than two minutes, and drilling fluid circulation is maintained at all times. It can operate with mud temperatures up to 250°F and pressures up to 17,000 psi.

Downhole motors

Unlike new bit types and MWD systems, downhole drilling motors are far from being new tools. But recent advances in the design of the downhole motor, coupled with new bits, promise to make the tool applicable to a much wider range of drilling conditions.

For many years, downhole drilling motors were used only to kick off a directional well. They are now being used increasingly to drill the straight portions of directional wells and to drill vertical holes.

Fig. 10-1 Downhole MWD assembly, left, and surface equipment, right. (courtesy Teleco Oilfield Services Inc.)

Until recent years, bits were not available that could withstand the high rotating speeds of a downhole motor. And some components of the motor were not durable enough, especially at high bottom-hole temperatures, to make use of the motors economical. Improved diamond bits and the introduction of the PDC bit will increase the use of downhole motors. Better materials for bearings and other components also have significantly increased the durability of today's motors.[8]

For instance, until recent years experts commonly felt that if a downhole motor could last for 100 hours, it would be economical. But in 1980, a downhole motor was used in an Oklahoma well for 429 hours, nearly 18 continuous days of drilling. In deeper holes, and as the cost of drill pipe increases, the use of downhole motors for drilling straight-hole intervals offers several advantages.

Because power is provided near the bit (Fig. 10-2) and not by the rotary table at the surface, drill pipe wear and the danger of failure are greatly reduced. The deeper the well, the longer the drill pipe must be rotated in the hole when the rotary table provides the power to turn the bit. With downhole motors, the bit is turned faster than in conventional rotary drilling, but the drill string can be rotated very slowly—just fast enough to prevent sticking the drill string. Also, the downhole motor often provides a more accurate surface indication of downhole conditions.

Downhole drilling motors are more efficient at transmitting energy to the rock. For example, friction losses from rotating the drill pipe and windup of the drill string reduce the amount of energy reaching the bit when the rotary table is used to turn the bit.

The two main types of downhole drilling motors are the positive-displacement type and the turbine type. Positive-displacement motors, in general, run at lower speeds and can be used with roller-cone bits. Turbine-type motors rotate at higher speeds and usually require the use of a diamond bit.

Downhole motors are powered by the flow of drilling fluid. The hydraulic horsepower of the drilling fluid stream is converted to mechanical horsepower to turn the bit. Bit speeds are higher and bit weight is lower than is the case with conventional rotary drilling. Depending on the type of downhole motor and its purpose, rotating speeds can range from 250 rpm to as high as 1,000 rpm. This means that the mud system on a rig using a downhole motor must supply more of the drilling energy than on a rig drilling conventionally. The rig's mud pumping capacity is a consideration when deciding whether or not to use a downhole drilling motor.

The development of advanced PDC bits along with continued improvement in downhole motor life will likely result in a combination (motor/PDC bit) that will significantly increase drilling speed and lower costs in many applications.

Deep drilling

The industry has long been concerned with developing equipment and methods to extend the depth to which wells can be drilled. Continued development will be aimed at further enhancing deep drilling capability, but not just to extend the depth that can be reached.

The industry has demonstrated that it can drill to depths of 30,000 ft and beyond. But the problems encountered in deep drilling are not related solely to depth. An 18,000-ft well in one area may pose more severe problems than a 25,000-ft well in another area.

The main factors that challenge deep drilling today are temperature, pressure, and the presence of acid gases.[9]

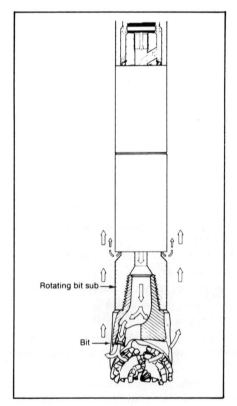

Fig. 10-2 Downhole drilling motor. (courtesy Dyna-Drill Division of Smith International Inc.)

The presence of acid gases—hydrogen sulfide (H_2S) in particular—means that all materials must be designed to prevent failure caused by hydrogen sulfide. When encountered at greater depths where temperature and pressure are higher, the presence of acid gas results in an even more severe environment. The development of higher-strength steels that will resist attack by H_2S will be needed, and better quality control and inspection methods will be required, according to one industry spokesman.[9]

Temperature limits most materials and equipment in deep drilling. It affects almost all components of a deep drilling operation, including the mud, tubular goods, logging tools, completion equipment, and stimulation and testing.

Temperatures that exist in deep wells, even without the presence of hydrogen sulfide, currently limit the use of many drilling fluids and drilling equipment. Extreme temperatures also call for improved seals and other components in packers, bits, downhole safety devices, and downhole motors.

Oil muds are more stable in high temperatures than water-base muds are, and they are used for drilling many deep wells. In deep, hot holes, oil-base muds also provide corrosion protection for the drill string and casing.

The use of oil muds is expected to expand for these and other reasons. But use could grow faster if there were better ways to remove drilled solids from an oil-base drilling fluid and to process the cuttings removed to make them more acceptable to the environment. Several companies are reportedly working on cuttings processing systems for oil-base muds that will make the cuttings environmentally compatible. Special systems have already been built and used successfully, but they have been expensive.

Efforts also continue to develop water-base muds that can withstand higher temperatures encountered in deep drilling. Many of the additives for water-base fluids available today also have severe temperature limitations.

Training

Rig crew training has advanced in recent years to a point where most rig personnel receive formal training using sophisticated techniques and training aids. Continued emphasis will be put on training in the years ahead.

Traditionally, much rig crew training was done on the job. A few special schools existed and a few training wells were available. But classroom training in blowout prevention and drilling operations left much to be desired because the student could not receive hands-on training. In addition, the number of students who could be trained on the few training wells was small.

The advent of electronic training simulators for the drilling industry began to remove these obstacles to adequate crew training. Originally, they were designed primarily to simulate loss of well control so rig crews could gain experience in handling well kicks. Now these simulators can pose a variety of drilling problems to the student and can represent most drilling operations.

Drilling conditions in the simulator can be changed easily by the instructor using a computer and a range of computer programs. Even the sounds common to a given rig operation can be reproduced by the simulator, adding to the realism of the training exercise.

Blowout prevention and kick control are still key goals in rig crew training, but the use of drilling simulators will also help rig crews learn to solve less-critical drilling problems and to drill more efficiently.

One recently introduced drilling simulator (Fig. 10–3) includes these components:[10]

Fig. 10-3 Drilling simulator. (courtesy Rediffusion Simulation Ltd.)

1. rig floor equipment
2. input/output system
3. surface and downhole mathematical models
4. computer system
5. instructor system
6. sound system
7. visual system

The unit can simulate a range of rig operations, including general drilling, casing and cementing techniques, mud control, and blowout prevention. A color graphic display of drawworks and rotary motion under student control is provided, and electronic sound generation provides synchronized background and operational noises.

The training complex includes generalized rig floor equipment, a choke manifold, drilling recorder/mud data equipment, and an instructor's station. The instructor monitors and reviews the training and can freeze exercises, speed them up or slow them down, and insert a wide range of malfunctions for the student to overcome.

Data gathering and analysis

The industry has developed sophisticated rig monitoring instrumentation and computer analysis techniques to evaluate logs and drilling data. To date, much of this technology has been limited to use on the most critical, most expensive wells. But the growing use of measurement-while-drilling equipment and other tools will bring more widespread implementation of the latest electronic monitoring and analysis systems, both on the rig and at remote sites where drilling information is needed quickly. Today's communications technology makes it possible to do almost anything with data gathered at the rig that is required for decision making or analysis.

An example of commercially available rig floor instrumentation is a system based on a self-prompted data processor that instantly provides up to 29 separate drilling parameters in digital or graph form. The unit presents data at a single source in common drilling terms and requires little training to use, according to the manufacturer.[11]

Its central processor, where program and data storage, data input, and system control functions are processed, can be set up at the rig site in an engineer's trailer, a mud logging trailer, or at the driller's control console. An engineer with a CRT can set alarm limits and check data independently from the driller. Hard copy in digital and graphic form can be provided by a printer, and additional CRTs can be installed at various rig locations if desired.

Another system recently introduced makes it possible for a drilling supervisor to monitor several rigs and to receive the same data available on the rig site with virtually no time delay at a central office any distance from the rigs (Fig. 10–4).[12] Rigs using the system transmit data via satellite to a central computer that sends the rig data to the supervisor's office. The computer stores all data so a supervisor can study a rig's past performance as well as current operating conditions. The supervisor can request broad overviews or detailed reports, he can set and monitor alarm conditions, and he can communicate directly with the rig via satellite.

The system consists of four main components:

1. the rig station, which includes equipment for collecting and transmitting data to the satellite, a CRT for displaying data, and a telephone/teletype arrangement for communicating with the supervisor
2. the satellite communications link
3. the earth station at Houston, which collects, formats, and stores all data while passing on specific information to the supervisor

Fig. 10–4 Central rig monitoring equipment. (courtesy Drilling Information Service Co.)

4. the drilling supervisory center in the supervisor's office, which contains an intelligent terminal with a CRT for data display, a line printer, and a telephone/teletype for communicating with the rig

Of course, few of the several thousand rigs drilling around the world have such sophisticated monitoring and analysis equipment. In many cases, such equipment cannot be economically justified. For instance, in a development drilling program where experience in a field has made it possible to define drilling conditions precisely, there is little need for such capability. But in a deep exploration well where unexpected conditions could be very costly, the latest monitoring equipment is often well worth the expense.

To some degree, this will always be the case. But the use of better data collecting and processing equipment will grow to keep pace with the need to cut drilling costs and to take full advantage of techniques such as measurement-while-drilling. During drilling, changes that must be made in the well plan as a result of unexpected conditions can also be determined more quickly and more accurately using computers.

Away from the rig site, the use of computers to help plan drilling operations will expand. The more data available for planning a difficult well, the better the chance it will be drilled without serious trouble. Analyzing these large volumes of data can be done with a computer.

Research needs

Spokesmen specializing in different phases of the drilling operation naturally have a variety of opinions on what the industry needs most to drill at less cost and more safely. For some idea of how industry professionals view those needs, however, the following discussion is based on a report published by the Bartlesville Energy Technology Center, U.S. Department of Energy, in April 1982.[13] It covers research projects that were identified at an industry-government workshop on Arctic, offshore, and drilling technology held at the Bartlesville Energy Technology Center, January 5–7, 1981.

Workshop participants identified 34 projects and six advanced drilling systems that should be investigated. These were separated into those that needed government research—seventeen projects and five advanced systems—and those that should be the responsibility of private industry. Individual projects are listed in Tables 10–1 and 10–2.

The projects the workshop participants felt would best be pursued by industry follow.

1. *Non-clay fluids*. At temperatures above 400°F most natural and synthetic polymers currently in use begin to fail to function as viscosifiers or filtrate control agents. A polymer is needed that has the structural strength to work either alone or with other polymers for this purpose at temperatures of 600–700°F.
2. *Cuttings cleaning (oil mud)*. Inadequate removal of oil from oil-mud cuttings has limited the use of oil muds, particularly in offshore drilling. Procedures to make oil-mud cuttings environmentally acceptable need to be developed or refined.
3. *Corrosion-related materials problems*. In deep drilling, the higher-strength steels needed are more susceptible to sulfide stress cracking, stress corrosion cracking, and corrosion fatigue cracking when H_2S is encountered. Materials are needed with yield strength, tensile strength, and bending or axial tension/tension fatigue strength at least 20% greater than those available today.

TABLE 10-1
Drilling research projects where government participation advised

Earth stress
Borehole stability
Post-well drilling data analysis for
    general correlations
Alternatives to strategic materials used
    in drilling for petroleum or other
    energy resources
High-temperature elastomers
Advanced directional drilling system
Rock properties from logs
Surveying accuracy
Jet cutting drilling
Fluid dynamics
Downhole/real time data collection and analysis
Horizontal drilling
Cementing high-angle holes
System integration analysis
Pore pressure and fracture gradient data analysis
Rock cutting phenomena
Improved seals
Drill-string analysis
Automatic drill pipe inspection
Coal seam sensor
Sensor concepts
Electrodrill
Magnetic anomaly data

Source: Reference 13

They also need corrosion rates at least 50% less than current materials.

4. *Solids control.* The industry needs more efficient methods to remove drilled solids from the drilling fluid and to mix mud materials and stir surface mud systems.
5. *High-performance lubricants for downhole tools.* The lubricants in journal bearing insert roller bits and sealed bearing roller reamers are adequate for conventional drilling; but at the higher speeds common with downhole motors, useful life is shortened. High temperatures involved in ultradeep drilling accentuate the effect of high speed. Lubricant materials need to be evaluated for service at rotating speeds of 600–1,000 rpm and temperatures greater than 200°C.
6. *Deep fluid motor for deep drilling.* Existing downhole motors use elastomers and seals that cannot withstand the high pressures (greater than 10,000 psi) and temperatures (greater than 200°C) encountered in deep holes. Existing mud motors and turbodrills need to be upgraded to handle these conditions.
7. *Downhole MWD/logging.* The workshop participants cited the need to develop specialized sensors to complement sensors currently under development.
8. *Directional drilling systems.* There is a need, according to the

TABLE 10-2
Drilling research projects best performed by private industry

Non-clay fluids
Cuttings cleaning (oil mud)
Corrosion-related materials problems
Solids control
High-performance lubricants for downhole tools
Deep fluid motor
Downhole MWD/logging
Directional drilling systems
Cuttings transport
Improved bits
Fluid abrasion
Oil/water separation
Drill pipe coatings
Production damage
Wellbore/cement/casing design
H_2S detection
Alternative cutting materials
Flexible drill string drill

Source: Reference 13

report, to "... coordinate and utilize developments in individual systems elements to produce an improved directional drilling system."

9. *Cuttings transport.* Cuttings transport in deviated holes is not easily predicted and often proves troublesome. Holes deviated at angles in the range of 40–50° give the most problems. There is a need for ways to transport cuttings out of high-angle holes at a more efficient rate without the need to increase the mud flow rate.
10. *Improved bits.* In citing the need to further the development of improved bits by continuing present research work, the report says reliable bits for hard rock, high-speed bits compatible with drilling turbines, and bits for directional drilling are needed.
11. *Fluid abrasion.* A trend toward larger-diameter holes and deeper drilling results in higher mud flow rates. Ways to reduce abrasion of internal surfaces of the drill string are needed. Using larger drill pipe and plastic coating internal surfaces are possible approaches.
12. *Oil/water separation.* Methods are needed that could reduce the oil content of water to below 50 parts/million (ppm) on a consistent basis.
13. *Drill pipe coatings.* Coating materials and application systems should be evaluated that will protect the drill pipe ID from chemical reaction with the drilling fluid while having no adverse effect on friction losses.
14. *Production damage.* Much effort is already being focused on ways to increase oil and gas production by reducing formation damage.
15. *Well bore/cement/casing design.* Serious casing stability problems result from not considering the well bore, the cement, and the casing as a unit, according to the report. Information, equipment, and procedures are needed to provide a permanent steel conduit from reservoir to surface.
16. *H_2S detection.* Hydrogen sulfide is currently detected by a gas chromatograph at ground level. Improvement is needed in the reliability of surface detection methods and the capability to analyze the severity of the gas more quickly, especially at greater depths. A sophisticated sensor in the drill string or the bit, coupled with a way to transmit the information quickly to the surface, would be desirable.

Only a few of these suggested areas for development involve new directions. Rather, they represent improvements that are needed in existing technology to meet the challenges of deeper drilling and more hostile environments. Those industry experts participating in this review of what is needed did not call for a basic study to find drilling methods to replace today's rotary system. This further emphasizes that the tools and techniques used today will be the heart of drilling operations for the foreseeable future. Steady improvement in materials and methods, rather than a completely new approach, will characterize oil and gas well drilling for the next two decades.

References

1. Slack, James B., and Jeffrey E. Wood. "Stratapax Bits Prove Economical in Austin Chalk." *Oil & Gas Journal*. (24 August 1981), p. 164.
2. Kelly, John, Jr. "Distinguished Author Series: Drilling Now." *Journal of Petroleum Technology*. (December 1981), p. 2293.
3. Johnson, V.E., Jr., et al. "Cavitating and Structured Jets for Mechanical Bits to Increase Drilling Rate." Paper presented at the ASME Energy Sources Technology Conference & Exhibition, March 7–10, 1982, New Orleans.
4. Moore, W.D., III. "MWD Will Change Drilling Techniques Over Next 5 Years." *Oil & Gas Journal*. (27 March 1978), p. 142.
5. McDonald, William J. "MWD: State of the Art—1: MWD Looks Best for Directional Work and Drilling Efficiency." *Oil & Gas Journal*. (27 March 1978), p. 141.
6. Moore, W.D., III. "Drilling Technology Will Play Bigger Role in 1979." *Oil & Gas Journal*. (18 September 1978), p. 137.
7. Product Data Brochure. "MWD Wireless Directional Surveys." Teleco Oilfield Services Inc., Meriden, Connecticut.
8. Luker, Stanley E. "Straight Hole Motors are Operational Reality." *Oil & Gas Journal*. (9 March 1981), p. 80.
9. Chadwick, Charles E. "Challenges of Deep Drilling—Conclusion: Mississippi Wildcat Shows Design, Planning Pay Off in Deep Drilling, Completing, Testing." *Oil & Gas Journal*. (2 November 1981), p. 102.
10. Product Data Brochure. "Advanced Digital Drilling Simulation." Rediffusion Simulation Ltd., Crawley, Sussex, England.
11. Product Data Brochure. "M/D 3200 Data Processing System." Martin-Decker, Santa Ana, California.
12. Product Data Brochure. Drilling Information Service Co. (DISC), Houston.
13. "Research Projects Needed for Expediting Development of Domestic Oil and Gas Resources through Arctic, Offshore, and Drilling Technology." Bartlesville Energy Technology Center, U.S. Department of Energy. (April 1982).

Index

A

abnormal pressures, 113, 114, 156
abrasion, 206
abrasive jetting, 4
AC power, 35
accumulator tanks, 166
acid frac treatment, 182
acid gas, 198, 199
acid treatment, 182
acoustic, 174, 195
acoustic velocity log, 176
AFE, 107
air compressors, 61
air-cushioned vehicle, 45
air drilling, 61, 62
air/foam, 61
air or mist, 62
aluminum drill pipe, 63, 139
American Petroleum Institute, 115
anchor chain, 131
anchor pattern, 131
anchoring system, 52
anchors, 131
angle buildup, 145
annular preventers, 162
annulus, 59, 119, 124
Arctic, 10, 45, 118, 139
areal extent, 188
articulated drill collars, 141
artificial lift, 184, 189
automatic station keeping, 42
auxiliary services, 50
axial load, 114

B

barge, 37
barge rigs, 39
barite, 60, 97
Bartlesville Energy Technology Center, 203
bauxite, 181
bearings, 70, 79
bent housing, 148
bent sub, 139, 147, 149
bit, 64, 149, 191, 199
bit balling, 73, 78
bit cost, 70, 127
bit damage, 79
bit design, 190
bit development, 74
bit, extended nozzle, 193
bit life, 69, 73, 77, 191
bit nozzle, 80, 92
bit, PCD, 191, 192
bit records, 81
bit selection, 81, 105
bit size, 112
bit weight, 73, 81, 191, 192
blind rams, 163
blowout, 81, 86, 98, 136, 143, 152, 155
blowout control equipment, 152, 160
blowout preventer, 36, 125, 161, 162, 164, 165, 167
blowout preventer drill, 166
blowout preventer stack, 164
blowout prevention, 153, 166, 200
bogie units, 45

borehole telemetry, 193
bottom-hole assembly, 140, 149, 150, 177
bottom-hole cleaning, 73
bottom-hole conditions, 193
bottom-hole pressure, 161
bottom-hole temperature, 85, 197
bottom plug, 125
bottom-supported rig, 31, 39, 40, 136
bottoms-up time, 90
box, 63
bulk cement, 123
bumper sub, 44
bumping the plug, 125
buoyancy modules, 134
burst, 115

C

cable tool rig, 2
calculated absolute open flow, 188
Campeche Sound, 169
cased-hole completion, 173, 179
casing, 65, 89, 109, 150, 206
casing cementing, 15
casing depths, 131
casing design, 105, 106, 107, 117
casing hanger, 110
casing joints, 115
casing packers, 180
casing point, 89
casing shoe, 125
casing size, 117
cavitation, 193
cement/cementing, 52, 65, 112, 121, 206
cement returns, 132
cement volume, 124
cementing casing, 15
cementing plugs, 124
cementing shoe, 121
center jet, 193
centralizers, 122
central platform/central pad, 137, 139
centrifuges, 98
chain storage, 131
chain/wire, 131
choke, 166
choke line, 134
choke manifold, 166
circulation, 121, 132, 161
coating materials, 206

collapse pressure, 114, 115
completion, 17, 137, 172, 187
completion equipment, 199
completion rig, 48, 172
compliant, 6
compressors, 85
computer analysis, 191
conductor casing, 132
cone, 70
continuous chain bit, 75
continuous phase, 93, 94
contractor, 13
contracts, 16
control system, 50
cooling, 86
coordinate system, 144
core, 77, 176
core analysis, 176
core barrel, 77, 176
core crusher bit, 73
coring, 77, 176
corrosion fatigue cracking, 204
corrosion protection, 102
cost estimate, 107
cost of drilling, 20, 69, 77, 83, 203
costs, other
    Arctic, 20
    average well, 23
    blowouts, 169, 170
    drainhole, 141
    drill pipe, 64
    floating rig, 8
    offshore drilling, 194
    offshore wells, 23
crew training, 199, 200
crooked hole, 73, 119, 135
crown block, 53, 54
curved conductor, 139
cuttings, 60, 84
cuttings analysis, 126
cuttings disposal, 95
cuttings processing systems, 199
cuttings transport, 206
cutting surface/cutting structure, 70, 79

D

Darcy, 174
data processor, 201
data transmission rate, 195

day rate, 16, 31
DC power, 35
deep drilling, 197, 198, 199
deep-water drilling, 5, 6
deep-water platform, 38
degasser, 127
depth, average well, 23
depth capability, 29, 62
depth, deepest well, 26
derrick, 15, 52
desander, 61, 98
desilter, 61, 98
development drilling/development well, 36, 105, 106, 155, 176, 194
deviated hole, 33, 64
deviation, 135, 148
diamond bit, 73, 79, 192, 197
diesel-electric rig, 35
diesel oil, 35, 94, 95
directional drilling, 15, 34, 136, 140, 143, 194
directional survey, 121, 140, 143, 145, 194, 196
directional well, 35, 135, 139, 145
diving systems, 52
dogleg, 151
downhole assembly, 195
downhole motors, 15, 34, 35, 57, 81, 139, 147, 148, 190, 192, 196, 197
downhole safety devices, 169, 199
downhole sensors, 195
drain hole, 141
drawworks, 10, 50, 52, 53, 54, 118
drill blanks, 191, 192
drill collars, 64, 92, 149
drilling break, 81, 127, 158, 159
drilling cost, 74, 102
drilling cost formula, 127
drilling data, 193
drilling efficiency, 193
drilling fluid, 36, 50, 61, 80, 83, 85, 86, 115, 155, 158, 190
drilling fluid density, 116
drilling fluid design, 105, 117
drilling fluid properties, 96
drilling fluid system, 59
drilling methods, novel, 67
drilling mud additives, 89
drilling plan, 154
drilling problems, 89

drilling/production platforms, 36, 38, 138
drilling rate, 116
drilling records, 107
drilling safety, 193
drilling sub, 195
drilling technology, 190
drilling vessels
  mobile drilling unit, 16
  drillship, 5, 7, 43
  dynamic positioning, 7
drill pipe/drill string, 53, 59, 62, 63, 125, 128, 149, 150, 163
drill pipe sticking, 150
drill pipe wear, 64
drill stem testing, 173, 177, 178
dropping angle, 145

E

efficiency, 68
electric arc drill, 4
electric log/electric logging, 174
electric rig, 55, 59
electromagnetic, 195
electron beam drill, 4
energy, 65
enhanced oil recovery, 142
entrained gas, 98
environment, 10
equations, 78
equipment, rig, 50
erosion, 63
excess cement, 124
exploratory well, 105, 106, 136, 155, 176
explosive drill, 4
extended nozzles, 193
extended reach drilling, 142, 143, 150
external pressure, 114

F

filter cake, 91, 92
filtrate, 97
fishing, 15, 52, 63, 65, 69, 79, 89, 130.
fishing tools, 130
fixed offshore platform, 6, 8, 36, 37, 131, 136, 137, 167, 194
flexible joint, 133
floating offshore drilling, 133, 161

floating offshore rig, 29, 31, 36, 39, 40, 43, 44, 131, 136, 137, 156, 157
flowing well, 112
flow tests, 182
fluid loss, 97
fluid saturation, 173, 174
foam, 83, 85, 91, 95
foam drilling, 62
footage rate, 16, 31
forces on casing, 114
forces on drill pipe, 63
formation characteristics, 173
formation damage, 102, 182, 187, 206
formation density log, 175
formation depth, 138
formation evaluation, 194
formation fluids, 155, 158
formation hardness, 79, 112, 149
formation pressure, 87, 95, 113, 114, 155, 156
formation types, 89
foundation pile, 132
fracture, 181, 182
fractured formations, 90, 142
fracture gradient, 114
free point, 92
free-standing riser, 134
fresh-water mud, 93

G

gamma ray log, 175
gas bubble, 157
gas-cut mud, 127, 158
gas kick, 157
gas lift, 185, 186
gas wells, 183
gel strength, 97
geology, 149
geophysical methods, 26
geopressures, 113
gravel packing, 180
guide lines, 132
Gulf of Mexico, 168, 170
gusher, 153
gyp muds, 94

H

hard formations, 78
hardwire, 195

heaviest string, 117
heaving/sloughing shale, 92, 94
helicopter-transportable rig, 45
high-angle wells, 33, 142, 143
high pressure, 110
high-pressure water jetting, 4
history, 1, 2, 3
hoisting capacity, 117
hoisting equipment, 57
hole angle, 135, 143, 145, 150
hole conditioning, 93, 121, 130
hole curvature, 112
hole deviation, 107, 119, 121, 135, 136, 138, 145, 150, 179
hole fillup, 149
hole inclination, 149
hole problems, 116
hole size, 112, 117
hole stability, 96
horizontal displacement, 138, 142
horizontal drain hole, 140, 142
hydraulic fracturing, 181
hydraulic horsepower, 197
hydraulics, 80, 148, 150, 192
hydrochloric acid, 182
hydrogen sulfide, 118, 159, 160, 199, 206
hydrostatic head, 87, 130
hydrostatic pressure, 87, 88, 113, 154, 157, 158, 159, 161, 177

I

ilmenite, 97
impermeable layer, 88
in gauge, 124
insert bit, 70, 79, 82
instrumentation, 201
intermediate string, 109, 112
internal combustion engine, 35
internal pressure, 114
International Association of Drilling Contractors, 20
intervals, 119
Ixtoc 1, 169

J

jackknife, 52
jackup rig, 7, 30, 39, 40
jet, 73, 80
jetting, 196

jetting bit, 148
jetting tool, 139, 148
joint, 121
journal bearing bit, 70
jungle areas, 11, 45, 119, 139
junk, 64, 69, 130

K

keeping the hole full, 130
kelly, 58, 160
kick, 61, 81, 167
kick control, 200
kick fluid, 161
kickoff point, 139
kill line, 134
kill mud, 161
kill procedure, 160
kill weight, 161

L

lag, 89, 193
land rig, 30
laser drill, 4
lease, oil and gas, 13
   cash bonus, 14
   competitive bidding, 14
   government-owned lands, 14
   offshore, 13
   simultaneous leasing, 14
lignosulfonate mud, 94
lime mud, 94
liner, 110
liner hanger, 110
liner/tieback combination, 110
log, 81, 173
logging, 15, 52, 65, 172, 173
logging tools, 199
log interpretation, 174
logging, real time, 194
logging systems, 194
log suite, 174, 194
lost circulation, 88, 110, 114, 119
lost-circulation materials, 88, 91
low-solids brines, 93
lubricants, 205
lubrication, 86

M

man-made islands, 10
marine riser, 36
mast, 10, 30, 52, 118
mat-supported jackup, 40
measurement while drilling, 128, 143, 193, 195, 201
mechanical drive rig, 35, 55, 59
mess facilities, 52
mill, 130
milled-tooth bit, 70, 79
minerals owner, 13
minimum yield strength, 115
mixing equipment, cement, 123
mobile rig/mobile drilling rig, 39
modules, 45
mud circulation, 81, 141, 146
mud cleaning equipment, 61
mud degasser, 98
mud mixing, 59
mud mixing equipment, 85
mud monitoring equipment, 126
mud motor, 149, 196
mud pit, 158
mud pressure pulse, 195
mud program, 107, 160
mud properties, 115
mud pump pressure, 50, 127
mud pumps, 10, 30, 59, 98, 118, 148, 158, 160
mud system, 30, 50, 115, 131, 155
mud tanks, 61
mud temperature, 196
mud treating, 59
mud weight, 80, 90, 159
multiple completion, 179, 183

N

natural gas/gasoline, 35
neutron log, 175
new bit, 127
normal pressures, 113

O

offshore blowouts, 152
offshore California, 167
offshore drilling, 5, 8, 117, 118, 136, 152, 156, 194

offshore fields, 138
offshore Louisiana, 168
offshore pipeline, 8
offshore platform, 139
offshore production, 8, 187
offshore rig, 30
offshore well, 118, 136, 194
oil-based drilling fluid, 93, 94, 183
oil mud, 192, 199
oil phase, 94
oil spills, 152
oil wells, 182
open-hole completion, 173, 177
operator, 13
Outer Continental Shelf, 168, 170
overbalance, 193
overpressured zones, 156
overpull, 53, 55, 117

P

packed-hole assembly, 150
packer, 177, 199
pad/drilling pad, 118
pendulum effect, 79, 145
penetration rate, 61, 73, 78, 80, 96, 116, 127, 141, 149, 159, 192
perforating, 173, 178, 179, 180
perforating tool, 179
permanent guide structure, 132
permeability, 159, 173, 176
permits, 13
pilot hole, 146
pin, 63
pipe inspection, 65
pipe ram, 163
pit level, 157
pit volume, 127, 158
plasma drill, 4
platform, 138
platform rigs, 37
plugged and abandoned, 12, 156
plugs, 91
polycrystalline diamond compacts, 73, 79
polymer, 204
porosity, 127, 173, 176
positive displacement motors, 197
potential test, 187, 188
power casing tongs, 121
pressure differential, 193
pressure gradient, 87, 156

pressure prediction, 113
price, oil and gas, 16, 17, 18
primary cement job, 126
producer, 173
producing equipment, 137, 172
producing life, 189
producing zone/formation, 110, 117, 172, 180, 181, 183
production casing, 173, 182
production string, 109, 110, 112, 173
productivity, 187, 188
proppant, 181
propulsion system, 42
protection string, 109
pump, beam, 183
pump, downhole, 183
pumping, rod, 183
pumping, submersible, 184
pumping, subsurface hydraulic, 184
pumping time, 124
pumping unit, 136
pumping well, 112
pump jack, 183

Q

quarters, 50

R

radioactive log, 175
radioactivity, 174
ram-type preventers, 162, 163
real time, 190
reamers/reaming, 65, 119, 180
regulations, 168, 187
relief wells, 136, 143, 144, 167, 169
research, 203
reserve pit, 61, 166
reservoir, 138, 173, 183
reservoir evaluation, 178
reservoir fluids, 158, 179
reservoir penetration, 138
resistivity, 174
rig building costs, 31
rig components, 117
rig costs, 30
rig crew, 15
rig data, 202
rig equipment, 95
rig monitoring, 201, 203

rig motion, 44
rig power, 35, 59
rig rating, 32
rig rental rates, 31, 32
rig selection, 106, 117
rig supply and demand, 30, 32
rig towing, 47
rig up/rig down, 55, 118
riser, 132, 133
riser connector, 132
riser system, 36
rock strength, 68
rod, 183, 184
rod pump, 136
rod string, 184
rolling cone bit, 70
rolling cutter rock bit, 3
rotary hose, 83
rotary rig activity, 17, 20
rotary rig/rotary drilling, 2, 30, 50, 83, 190, 196
rotary table, 10, 50, 57, 58, 128
rotating drilling head, 61
rotating speed, 50, 69, 77, 80, 81, 82, 145, 150, 192
royalty, 13, 14
Russia, 26, 34

S

S curve, 144
safety, 62
saltwater muds, 94
sand control, 180
Santa Barbara Channel, 167
scratchers, 122
sea conditions, 44
sealed bearing, 70
seawater mud, 93
self-potential, 174
self-propelled offshore rig, 131
semisubmersible, 7, 31, 40
sensors, 195, 205
setting depth, 110, 113
shale shaker, 61, 84, 98
shallow fields, 33
shear rams, 164
sidetrack, 64, 126, 146, 156
sidewall coring, 176
silicon-controlled rectifier, 35, 36
simulator, electronic, 166, 191, 200

single, 53
single completion, 179
single curve, 144
slant rig, 33, 139
slip joint, 44, 133
slips, 58, 128
sloughing shale, 119
soft formations, 78
solids accumulation, 130
solids content, 97, 127
solids control, 96
solids removal equipment, 84, 117
sonic log, 176
specialty rigs, 34
spill containment, 169
spontaneous potential, 174
spotting oil, 92
squeeze cementing, 126
stabilizer, 65, 141, 149
stable foam, 95
stand, 53, 128
standard derrick, 32
standpipe, 83
steam-powered rigs, 35
steel-tooth bit, 70
stiff foam, 95
stimulation, 173, 181, 182, 199
storage, 50
straight-hole intervals, 197
Stratapax, 191
stress corrosion cracking, 204
stuck pipe, 91, 92, 156
sub, 149
submersible rig, 39
subsea blowout preventer stack, 132
subsea completion, 7, 143
substructure, 53
sulfide stress cracking, 204
support equipment and services, 15
surface owner, 13
surface string, 109
survey tools, 193
swabbing, 155
swivel, 83

T

telemetry system, 194
telescoping joint, 133
temperature, 26, 94, 198
temperature gradient, 85, 86

temperature log, 125
temperature stability, 102
temporary guide base, 132
tender, 37
tension, 114
tension leg, 6
tensioning system, 44, 134
test equipment, 187, 189
testing, 173
test separator, 187
three-cone bit, 148
thrusters, 7, 42
tieback string, 110
tool joint, 63, 164
top plug, 125
torque, 63
training, 191
training wells, 166
traveling block, 54
treating equipment, 86
trip, 53, 59, 69, 70, 155
trip cost, 127
triple, 53
trip time, 191
truck-mounted rig, 33, 45
tubing, 182, 183
tubular goods, 199
tungsten carbide, 70, 191
turbines, 196
turbodrills, 205
turnkey contract, 31
twistoff, 63
two-cone bit, 73
types of rigs, 30

U

U.S Department of Energy, 203
U.S. Geological Survey, 170

V

valves, downhole safety, 186
valves, gas lift, 186
vertical holes, 196

vertical well, 135
vibrating screen, 61
viscosity, 60, 92, 97

W

waiting on cement, 119
washouts, 124
water-based fluid, 93, 199
water depth, 29, 37, 40
weight material, 60, 116
weight of casing, 53
weight on bit, 50, 69, 77, 79, 80, 81, 82, 145, 150
well angle, 138
wellbore stabilization, 100
well completion, 47, 48, 172
well control, 86, 116, 127, 152, 153, 156
well control equipment, 44, 106, 170
well control procedures, 158, 167
well control training, 166, 170
well cost, 70, 104
wellheads, 137
well kick, 61, 157, 160, 200
well log, 106, 124
well owner, 13
well path, 145
well planning, 105, 152
well potential, 173
well pressure, 152
well records, 81
well servicing rig, 47
well treatment, 181
well types, 11, 12, 23, 26
well workover, 47
whipstock, 140, 145, 146
wildcat, 81
wireline logging, 176
workover rig, 47
worn bit, 127

Y

yield point, 97